# THE BODY AND SOCIETY

# Theory, Culture & Society

*Theory, Culture & Society* caters for the resurgence of interest in culture within contemporary social science and the humanities. Building on the heritage of classical social theory, the book series examines ways in which this tradition has been reshaped by a new generation of theorists. It will also publish theoretically informed analyses of everyday life, popular culture, and new intellectual movements.

# THE BODY AND SOCIETY

## Explorations in Social Theory

### *Second Edition*

BRYAN S. TURNER

**SAGE Publications**
London • Thousand Oaks • New Delhi

© Bryan S. Turner 1996

Second edition published by Sage Publications 1996

First edition published by Basil Blackwell Publisher 1984

 SAGE Publications Ltd
6 Bonhill Street
London EC2A 4PU

SAGE Publications Inc
2455 Teller Road
Thousand Oaks, California 91320

SAGE Publications India Pvt Ltd
32, M-Block Market
Greater Kailash – I
New Delhi 110 048

Published in association with *Theory, Culture & Society*,
Nottingham Trent University

**British Library Cataloguing in Publication data**
A catalogue record for this book is available
from the British Library

ISBN 0 8039 8808 7
ISBN 0 8039 8809 5 (pbk)

**Library of Congress catalog card number   96–68999**

Typeset by M Rules
Printed in Great Britain by The Cromwell Press Ltd,
Broughton Gifford, Melksham, Wiltshire

to ei

# Contents

# Acknowledgements

My interest in developing a genuine sociology of the body emerged out of an earlier interest in the sociology of religion which is clearly evident in my *Religion and Social Theory*, a volume which was in turn intellectually stimulated by the late Trevor Ling and by Roland Robertson. In more recent years, my understanding of the theme of embodiment has been enhanced by working with Mike Featherstone, Chris Rojek, Pasi Falk, and Gary Albrecht. At Deakin University, my colleagues in the Centre for the Body and Society have been a source of daily inspiration. I owe a special thanks to Anne Riggs, my co-researcher, for her work on intimacy and ageing, and generations and life-style. In terms of their editorial support, Robert Rojek and Stephen Barr have been consistently generous. Finally, Eileen has been the eye to my soul.

A version of Chapter 3 first appeared in the *Annual Review of the Social Sciences of Religion* (1980) 4, pp. 247–86.

# PREFACE TO THE SECOND EDITION

# REPRESENTING THE BODY AS A PICTORIAL METAPHOR

With the publication of the second edition of *The Body and Society*, it is timely to reflect upon the general problem of the artistic representation of the human body and embodiment in the humanities and the social sciences. The first edition of *The Body and Society*, which appeared in 1984, utilized Pablo Picasso's 'Les Demoiselles d'Avignon' as the jacket cover design. This painting from 1907, which expressed Picasso's interest in African masks, was suitable in expressing the problematic status and nature of the human body. The mask-like faces of the five principal characters of the painting summarized much of the ambiguity of the relationship between gender and sex, the relationship between symbol and body, and the interaction between western and non-western representations of the female figure. The painting also illustrates a brothel scene in France, but the women's bodies are so abstract that it is difficult to characterize them easily as either masks, human effigies, real bodies, or symbolic marks of female subordinate status. In one sense, Picasso's design enabled the book to get away from the problem of representation by simply asking questions about what is being represented.

In retrospect this painting by Picasso was too obvious or too literal as an expression of the problem of the body and embodiment. By 'knowing' that the painting was by Picasso, the viewer/reader can locate these symbols within a well-known rhetoric of modern art, concerning female sexuality, French colonization of African culture, desire and image. In fact most recent jacket designs for sociology of the body have also adopted rather conventional approaches to the issue. For example, Chris Shilling's *The Body and Social Theory* employs Fernand Léger's 'Deux Femmes avec des fleurs' as a design but the statement it presents is relatively obvious in the sense that the images clearly convey a message about sexual and gender identity through a painting which clearly represents the female body as a sexual figure, for example through the female breast. John O'Neill's *Five Bodies* represents the human shape of modern society via a version of Leonardo's sketch of the perfect male body. Emily Martin's *The Woman in the Body* again returns to Pablo Picasso in order to use a 'readable' illustration. Picasso's 'Girl Before a Mirror' from 1932 is an illustration which, again in terms of a set of abstract shapes, clearly depicts a female body. Arthur W. Frank's recent *The Wounded Storyteller: Body, Illness and Ethics* adopts a sensuous black and white photograph of a male torso in order to address this question of representation. Drew Leder's philosophical inquiry into *The Absent Body* refers to Magritte's 'Le Pèlerin' of 1966 to make a statement about absent faces as a sign of the absent body. Perhaps even more conventionally Peter Brown in *The Body and Society* reiterates the Adam and Eve theme to write about sexual renunciation in early Christianity. On the basis of these cover designs[1] one could conclude that representing the body in the social sciences and humanities involves adopting a set of well-known conventions, which grapple with gender images, social conventions surrounding what can be represented or spoken about, and what can be understood.

One of the specific issues in representations of the body is the question of the relationship between gender and sex in a context of both feminist and queer politics which have challenged conventional or traditional understandings of the relationship between anatomy and social roles. As a consequence of the politicization of the image of the body in the post-war period, it is difficult to select a neutral image of the body which can convey a general message about politics and embodiment, rather than a specific statement about sex and politics. In fact the point of feminist and queer politics would be to suggest that there is no such thing as a neutral representation of gender and sex. Representations are not innocent; they carry, for one thing, an irredeemable cultural history of sexual relationships. In the visual culture of embodiment, it is easy to be seduced by images of strong bodies, sporting bodies, flowing forms, rounded shapes and the muscular or fluid torso as the representation of living sexuality.

In this second edition of *The Body and Society*, Francisco Goya's still life of the head and ribs of a sheep is used as the cover illustration for the book. Francisco José Goya y Lucientes (1746–1828) seems a particularly useful and suitable artist to illustrate a sociological discussion of embodiment. He spent the first part of his life in the peaceful environment of the rule of Charles III, but lived the remainder of his life in the turbulent atmosphere of social unrest which followed the French Revolution. His enormous output of paintings reflect the passion of his life's work. Goya was very heavily involved in the court-life of Madrid and in addition to his court scenes illustrated central aspects of Spanish social experience through, for example, the bull ring, peasant dancers, the mad-house and revolutionary politics. As the result of illness, he was left completely deaf at the age of forty-seven. His court paintings in particular reflected an elite life-style characterized by cruelty, despotic violence and voluptuous indulgence. The portraits of Ferdinand VII and Queen Maria Luisa are famous as representations of power, banality and sensuality. In representing these themes, he developed a particular methodology in which his paintings represent social life through density rather than through lines.

In the history of art, representations of animal flesh have a poignant and melancholic role in their relationship to the artist and audience. One can think about Rembrandt's representations of the ox carcass, which were movingly adopted by painters like Chaim Soutine, as a reflection upon human marginality and death. In fact dead meat perfectly indicates the complex status of embodiment and the body, regardless of whether we are talking about human or animal existence. Meat is beautifully (perhaps one should say, hideously) located mid-way between nature and society, between nature and culture, between the living and the dead. From an anthropological point of view, it is a culturally transformed food on its way to consumption to sustain human bodies which in turn are subject to the processes of death and decay. This picture of meat is a shocking representation of the major questions about cultural processes in relation to the human body, the status of food, the question of animal rights, the relationship between the human body, history and meat. *The Body and Society* is in fact a sustained question about the nature of the body, a topic which is rejected in favour of the notion of embodiment. The concept of the 'body' suggests a reified object of analysis, whereas 'embodiment' more adequately captures the notions of making and doing the work of bodies – of becoming a body in social space. Dead meat 'ideally' raises questions about our embodiment, about our frailty and our destiny, about the relations between luck and ethics. Furthermore our gaze over this flesh raises questions about the pornographic status of dead meat in human societies. The still life is redolent of sexual emotion and death. It suggests to the viewer a pornographic relationship to food as the basis of life and community. It brings into question the place of the mouth within the body and within culture. The theme of the butcher's table in Goya's work implicitly indicates the idea of sinking teeth into flesh. Teeth are at once part of the practices of sexual pleasures, and of death and violence. The death of this animal becomes the basis for human survival, for the human species as carnivorous beasts. The mouth as body orifice is also an organ of speech, consumption and violence. We appropriate the world through the mouth, as our original social link with our mothers, as an organ of speech and articulation, as an organ of consumption and animal violence. Eating is the origin of community, where festivals are celebrations of belonging and membership through a sharing of food. The sharing of bread or com(pan)ionship is the basis of the social.

Meat also stands for my problematic existence within a theological context as flesh, as death, as sinfulness. Dead meat indicates the homelessness of human kind, our problematic status with respect to the very processes of life and existence, and our alienation from our own bodies. The sharing of bodies or The Body became the basis of Christian community in the injunction 'Take, eat!'

*The Body and Society* is centrally concerned with questions about disease, death, nature, corruption and the idea of the human body via Christian theology as flesh. The idea of a carcass is thematically very close to the central issues of this book, which in discussing Christian theology notices the historical conceptualization of the body as merely flesh, as a dead animal. The painting of the ox carcass in Rembrandt, Soutine and Goya is full of symbolic indicators about the frailty of human existence, the brutality of life as a consequence of embodiment, the process of ageing, the fragility of the human being despite our power over animals and the complex relationship between eating, existing, food, sex and the body. This picture directly addresses us with a question: who are we? More profoundly, it confronts us with the interrogation: what are we? Again from a religious perspective, the relationship between production and reproduction, between sacred and profane, is raised through pastoral images of the connection between animals and human beings. The totemic animal is a crucial figure in the mythologies of 'primitive religions'; their sacrificial slaughter made society possible. Christianity evolved out of these sacred narratives. Christ is the sacrificial gift who makes our salvation possible. Christ as the Good Shepherd gathers his Flock together as a Church, but it was the Good Shepherd, not the sinful sheep, whose body is consumed by his Flock in a eucharistic meal. Images of slain sheep carry a religious message, despite the overtly secular context of the butcher's table. In this picture of the slaughtered sheep, the animal now rests at ease in a status of death, but transformed by the cultural practice of the still life into an object of beauty, an object of aesthetic contemplation. These images of the body as animal flesh, as artistic representation and as a forgotten narrative of religious exchange indicate the proximity of the body to metaphors of sociality. The Church is the body of Christ, the Good Shepherd of wayward Sheep. The sharing of bread (pan) provided a discourse for all forms of companionship and community. The process of eating is thus transcribed into a discourse of social relationships and exchanges. In contemporary society, with rationalization, secularization and McDonaldization, these robust metaphors of body as centrepiece of human thinking are now disguised, submerged, or displaced by technology. It is often through the history of art, for example, that we can grasp dimly these body metaphors which lie, but no longer to hand, at the foundation of thinking and feeling. Goya's still life of the carcass of a sheep conjures up, if we will let it, a prehistory of metaphors of existence. At the same time, this conjurement interpellates identities which, through a system of potent meanings, shape the very fabric of society.

*Bryan S. Turner*

## Note

1   These book covers are from Chris Shilling (1993) *The Body and Social Theory*, London; John O'Neill (1985) *Five Bodies: The Human Shape of Modern Society*, Ithaca and London; Emily Martin (1989) *The Woman in the Body*, Milton Keynes; A.W. Frank (1995) *The Wounded Storyteller: Body, Illness and Ethics*, Chicago and London; Drew Leder (1990) *The Absent Body*, Chicago and London; and Peter Brown (1988) *The Body and Society: Men, Women and Sexual Renunciation in Early Christianity*, London and Boston, Mass.

# THE BODY AND SOCIETY

# INTRODUCTION TO THE
# SECOND EDITION

# THE EMBODIMENT OF
# SOCIAL THEORY

When *The Body and Society* was first published in 1984, there was little interest in mainstream social sciences and humanities in the sociology of the body
apart from a number of publications which had been influenced by the work
of Michel Foucault. In the intervening decade there has been a flood of publications concerned with the relationship between the body and society, the
issue of embodiment with relation to theories of social action, the body and
feminist theory, and the body in consumer culture. These developments have
been reviewed extensively in a number of publications (Turner, 1991; 1994;
1995), but in this new introduction to *The Body and Society* my aim is to
bring about a consolidation of contemporary thinking about the sociology of
the body and to identify important lines of development for research and
analysis. In particular, I analyse the social changes which have given a special
prominence to the body in contemporary social theory and review important
historical changes in the presence of the body in society, taking special notice
of the relationship between religion and the body. I have attempted to capture
this historical dimension to the body in society through the notion of a
'somatic society' (Turner, 1992), namely a society within which major political and personal problems are both problematized in the body and expressed
through it.

This introduction outlines some of the major theoretical issues which are
faced by the attempt to produce a sociology of the body, paying particular
attention to the so-called social constructionist debate which has been important, for example, in medical sociology and feminist theory. I suggest some
theoretical consolidation of these various approaches, attempting to pull
together many of the diverse and conflicting strands in contemporary
approaches to the body in the humanities and social sciences. Finally, an outline of a research agenda is presented which would drive the sociology of the
body and related areas into some integrated and productive empirical research.
Theoretical and research development require a certain consolidation of conceptual and theoretical approaches, and this objective may be reinforced by
better appreciation and understanding of the legacy of social anthropology
and its contribution to an understanding of human embodiment. A successful

programme of development for the sociology of the body, as with other components of contemporary sociology, requires an empirical focus, a research agenda, a commitment to a political perspective on the application of sociology, and an infrastructure of research, such as journals, publication series and international conferences (Turner, 1996).

## Why the Body?

This introduction is based upon the assumption that our current interest in and understanding of the body are consequences of the profound and long term transformation of Western industrial societies. In particular the prominence and pervasiveness of images of the body in popular and consumer culture are cultural effects of the separation of the body (and in particular its reproductive capacities) from the economic and political structure of society. The emphases on pleasure, desire, difference and playfulness which are features of contemporary consumerism are part of a cultural environment which has been brought about by a number of related processes namely postindustrialism, postfordism and postmodernism. The moral apparatus of bourgeois industrial capitalism with its religious and ethical condemnation of sexual enjoyment has largely disappeared with the erosion of Christian puritanical orthodoxy and the spread of mass consumerism. These changes in the moral and legal apparatus of late industrial society are in turn associated with changes in the structure of the economy, particularly the decline of heavy industrial production for a world economic system. The increasing importance of service industries in a postindustrial environment has been associated with a decline in the traditional urban working class, and with life-style changes associated with unemployment, early retirement and increasing opportunities for leisure. Economic change and restructuring have brought about fundamental shifts in the nature of labour and its composition, which have also reorganized leisure and consumption. Young, working-class men have become a surplus population, whose machismo image of toughness no longer has a direct functional relevance to their enforced leisure. The labouring body has become a desiring body.

   These social and cultural changes have to be seen within a broad historical context, namely the decline of a feudal system based upon the ownership of land, the rise of industrial capitalism based upon the control of industrial processes and finally the emergence of a postindustrial or postmodern society organized around the control of communications and sign systems. In traditional societies there was a close structural relationship between the ownership and reproduction of property and the ownership and reproduction of human beings through a patriarchal monopoly of women within an extended family. Marriage and the ownership of domestic property were crucial institutions in this nexus between the ownership of property and the ownership of bodies. Men within a patriarchal household controlled the inheritance and distribution of property (particularly land) through the control and ownership of women as reproducers of men. With the decline in the importance of land in the production processes of industrial capitalism, the problem of control and ownership shifted to the capitalist family where industrial wealth was inherited through the male line. Although men in Victorian Britain, for example, enjoyed the benefits of a double standard, the importance of traditional sexual mores in the household was emphasized because of their crucial relationship to the inheritance of wealth. These familial and economic systems explain the importance of virginity, fidelity and sexual purity in this period.

With the movement of industrial capitalism towards a postindustrial system based upon a global economy, service industries, advertising and advanced consumption and the manipulation and control of communications through public relations industries, this traditional relationship between property, sexuality and the body has largely disappeared. The erosion of this system has been marked by the growth of divorce legislation making conjugal separation and divorce largely independent of material grievance or harm. Divorce on demand has signalled the arrival of a new period of serial monogamy, fragmented life patterns and diversified life careers. The availability of mass, cheap contraceptive means has transformed interpersonal relations, offering in principle the possibility of a wide range of partners and relationships. These technical and legal changes in contraception are also closely interrelated with certain demographic changes in the populations of the advanced societies, namely a decline in the birth rate and an increasing expectation of longevity.

These socio-economic changes are furthermore closely associated with what Anthony Giddens has called *The Transformation of Intimacy* (1992), in which self-understanding, individualism and self-realization are expressed through interpersonal relations based upon pure emotion, non-utilitarian trust and interpersonal intimacy. As Giddens argues, personal and domestic relations are no longer based upon a property contract but on a series of expectations about personal satisfaction through intimacy and sexual contact. These new relations of intimacy also express significant changes in the nature of identity and personhood in postindustrial societies (Giddens, 1991). The body, as we will see, is crucial to these new patterns of expressivity and intimacy because the body is the channel or carrier of these new emotional intensities.

The consequences of a postindustrial culture are important for sociological understanding of consumerism, commercialization and hedonism. There is obviously a strong commercial and consumerist interest in the body as a sign of the good life and an indicator of cultural capital. In addition to this theme of consumption there is a specific focus on the body beautiful, on the denial of the ageing body, on the rejection of death, on the importance of sport and on the general moral value of keeping fit. In early capitalism there was a close connection (an 'elective affinity' as Max Weber described it) between discipline, asceticism, the body and capitalist production, while in late capitalism there is an entirely different and corrosive emphasis on hedonism, desire and enjoyment. These ideas provided the basis for Daniel Bell's analysis of *The Cultural Contradictions of Capitalism* (Bell, 1976). Given the emphasis on leisure, individual expressivity and consumption, the body emerges as a field of hedonistic practices and desire in a culture which recognizes that the body is a project. There are, of course, important social variations on this theme, particularly in terms of social class, gender and generations. In contemporary youth culture where there is a chronic problem of unemployment, the body is pierced, decorated and tattooed as a collective symbol of tribal belonging. However, these patterns of membership, while intense and affective, are episodic, transitory and fragmented, resulting in neo-tribalism (Maffesoli, 1995). The body has emerged into theoretical and political prominence because of these major transformations and the underlying nature of the economy, property and employment.

The body has also come to prominence because of the political impact of the women's movement, feminist criticism of patriarchal social organization, and the transformation of the role of women in the public sphere. It has been suggested plausibly that the neglect of the body in classical sociology was brought about by the fact that sociology has been largely male sociology (Frank, 1991; Seidman, 1994). One obvious objection to this claim would be that anthropology has maintained a consistent focus on the relationship between body and culture in, for example, the well-established tradition of research into rites of passage. Anthropology like sociology has been dominated by men. It is hardly surprising, given the impact of the women's movement, that there has been a rich development of feminist theories of the body and embodiment (Grosz, 1994), especially in the work of Julia Kristeva and Luce Irigaray.

The women's movement is in fact a political manifestation of the underlying economic and social changes taking place in the economy and the family. In many paradoxical ways the emancipation of women was closely associated with their introduction into the male labour force as a consequence of the emergence resulting from the First and Second World Wars. As women entered the male work-place and became industrial labourers, they shed many of their previous attitudes towards family life, children and the subordinate nature of female labour in traditional societies. As we have seen, greater access to contraception and the availability of divorce through more liberal legislative means resulted in a more assertive and independent women's movement.

While many of the benefits of domestic technology and contraception have been challenged and criticized by modern feminism, they did contribute to a profound change in the status of women in the society. This changing political status has been expressed through various feminist debates about the problematic nature of gender, sex and sexuality. This questioning of the given nature of sex differences by anatomy and biology resulted in a significant social constructionist critique of traditional notions of sexual difference and diversity. As a result feminist theorists have questioned the foundational nature of sexual difference asserting that the differences between men and women are historical, cultural and contingent rather than fixed by nature and divine will.

Many contemporary writers on the body have claimed that the body is now part of a self project within which individuals express their own personal emotional needs through constructing their own bodies (Shilling, 1993). Given the emphasis on selfhood in contemporary consumer culture, the body is regarded as a changeable form of existence which can be shaped and which is malleable to individual needs and desires. Of course, in one sense this is now literally true in that contemporary microsurgery has made transsexualism an available option for a proportion of the population (Lewins, 1995). Both feminist and gay literature therefore tends to adopt an anti-foundationalist view of the body, emphasizing instead the changeable, malleable and contingent characteristics of embodiment in modern societies.

The body has become important in contemporary culture as a consequence of major changes in the nature of medical practice, medical technology, the changing structure of disease and illness, and finally because of the greying of the human population, at least in the advanced industrial societies. The ageing of the population is a significant political and social issue for at least two major reasons. First the characteristics and prevalence of modern forms of degenerative disease are obviously closely related to the ageing of contemporary populations. For example, in contemporary industrial societies there has been a significant decline in the prevalence of infectious diseases since the Second World War. The decline in the infantile mortality rate and improvements in general life expectancy have contributed to the greying of populations and in turn this has resulted in a new range of 'killer diseases'; the principal causes of death are now diseases of the circulatory systems, the respiratory system and from malignant neoplasms. Secondly, the ageing of the population of industrial societies has had a significant social impact on the economic performance of capitalism because of the increase in the dependent populations associated with ageing and retirement. Industrial societies will increasingly face a situation where significant proportions of their population are retired, elderly or disabled, and these demographic changes will have an important impact, not only on economic productivity but on patterns of consumption, leisure and lifestyle.

These changes in demographic structures of societies, in life expectancy and in the balance between the sexes should be seen within a larger framework of social change which includes such issues as artificial insemination, in vitro fertilization, new reproductive technologies, a global organ transplant industry, heart transplants, the growth of cybernetics, microsurgery and other advances in pharmacology. The impact of these scientific changes associated with new medical technologies have raised major philosophical, ethical and legal issues in contemporary society which ultimately are related to the nature of personhood, identity and individualism. These social and scientific changes have brought into the question the whole basis of the legal ownership of parts of human bodies, particularly reproductive material. Furthermore this new environment of risk questions the role of the state in protecting the sick, the elderly and disabled from unwarranted and undesirable medical experimentation and intervention. These medical changes have raised many new questions about the traditional philosophical dilemmas surrounding the relationship between the body and the soul, embodiment and the self, consciousness and identity and so forth. Government responses to the epidemic of HIV and AIDS have also emphasized the problematic nature of medical intervention with respect to individual rights and liberties. These changes and dilemmas are all part of the general background for what I have called the rise of the somatic society, a society in which our major political and moral problems are expressed through the conduit of the human body.

The human body therefore has become a focus of much social science and humanities research as a consequence of these macro changes in the social, economic and legal status of human embodiment in a society of rapidly expanding technology. Whether or not these changes bring about an expansion of the normalization and rationalization of the human body as suggested by the work of both Weber and Foucault (Turner, 1992), or whether, for example, cybernetics offer new forms of liberties and social change in terms of the work of radical feminists like Donna Haraway (1985; 1991) is an issue much debated in the literature. The information superhighway and virtual reality will bring about new ways of conceptualizing the body in relation to time and space, offering both new opportunities for democratization and authoritarian control of the human self as an embodied agent. Alongside these social and political changes there have as a consequence been major developments in social science fields related to the body, for example, in the whole construction of gerontology (Green, 1993), governmentality (Burchell, Gordon and Miller, 1991), women's studies, the analysis of masculinities and the anthropology of nutrition and ethnomusicology.

**The Body in History**

The body has become somewhat disembedded from many of the major institutions of contemporary society, particularly in terms of its relationship to the family, reproduction and property ownership. The body no longer functions within the crucial interplay of property, wealth and inheritance in the household economy; it is no longer so clearly a focus of marriage strategies, monarchical debates or inter-state violence as symbolized by the violent conflict of heroes. This social dislocation of the body means that the body has been more exposed to the playful manipulation of consumerist culture becoming a principal vehicle for what one might call consumerist desire. This separation of the body from its traditional sexual functions is associated with a women's movement which has questioned the relationship between sex and gender, and a gay movement which has problematized the nature of the masculine body.

This separation of the body from many traditional functions has also resulted in a transformation of the location of the body with respect to eating, eating rituals and food ways. The emergence of the modern self is closely related to the development of consumerism and consumption. Following the research of Colin Campbell (1987), Pasi Falk claims that the modern sense of the self is significantly connected with the notion of unlimited personal consumption of consumables and 'pleasurables' such as food, signs and consumer goods. We might argue, converting Descartes's famous individualistic slogan of the cognitive self, that 'I consume therefore I am'. The notion of a consuming self has the ideal character of the modern individual as being embraced and amplified by the advertising and public relations industries.

In this regard Pasi Falk also offers us a significant critique of Foucault's treatment of the self as merely the effect of modern practices of discipline and technology. Foucault's continuing commitment to structuralism has subordinated not only the phenomenology of embodied experience, but also removed any capacity for resistance or opposition to disciplinary regulations and practices. More exactly we should argue that the notion of the self in consumer society ought to be seen in terms of the body-image that plays the distinctive role in the understanding and evaluation of the self within the public arena (Schilder, 1964). It is typically the surface of the body which is the focus of advertising, self promotion and public relations. It is also the body surfaces which are the foci of the social stigmatization. As I argued in the first edition of *Body and Society*, the modern consuming self is a representational being, as understood in the work of Erving Goffman.

In order to understand the location of the body within contemporary society we need a better grasp of the historical development of the body in society. In this introduction I shall concentrate primarily on the question of the body in Western history drawing attention in particular to the status of the body in relation to the traditional dichotomy between the sacred and the profane. The human body is central to both religious and secular mythologies because it provides an obvious foundation for metaphors and myths for thinking about the individual and social relations. It is the foundation in fact of mythological systems at least from an anthropological point of view. Mary Douglas in *Purity and Danger* (1966) and *Natural Symbols* (1970) has analysed the importance of body metaphors in the analysis of risk, ambiguity and uncertainty within social systems and for the individual. These metaphors of risk and danger are characteristically associated with the orifices of the human body. For example, many of the New Testament images of Christ's body are concerned with the question of purity and danger with respect to, for example, the flow of blood from the broken body of Christ from the cross.

In addition to body orifices there has been a long tradition of associating sacred and profane attributes with the right and left-handed characteristics of the human body which is universally asymmetric. Some of these issues regarding handedness as a classificatory system were raised by Robert Hertz in his *Death and the Right Hand* (Hertz, 1960). This essay by Hertz, which followed the tradition of Durkheim and Mauss (1963) has been influential in the development of anthropological theories of classificatory systems based upon a binary opposition. For Hertz the pre-eminence of the right hand in symbolic systems is a social institution, or in Durkheim's terminology a social fact, because the use of the right hand is positively sanctioned by human society and typically deviations from right-handedness are punished. Hertz regarded the polarity between right and left as a basic expression of the religious dualism between the sacred and profane which had been basic to Durkheim's approach to the sociology of religion. The right hand represents the sacred side, male values, life, virility and power. While the left hand stands for evil, death and the female. In heraldry the left side design is regarded as the 'bend sinister', which produced the notion of sinister. The sinister bend stands for illegitimacy. In contemporary society, of course, right-handedness is still associated with worthiness and with ideas related to dexterity and beauty. Thus the body offers a profound and rich source of metaphors and similes and modes of conceptualization for the crises, hazards, dangers and paradoxes of individual and collective existence.

The relationship between divine bodies, human bodies and social bodies is another dimension of human mythology. In particular we can think about the materialist characteristics of Christian mythology and its emphasis on Christ as the incarnated, suffering, divinity who created a system of social solidarity with human beings through the ritualistic consequences of the sacrificial Eucharistic meal. In Pasi Falk's study of *The Consuming Body* (1994) the question of orality is examined with respect to types of solidarity through communal meals and other forms of exchange in rituals, social gatherings and collective festivities. The Church was also seen as the body of Christ, namely as a corporation which embodies the authority of a charismatic Christ through the Church's 'Keys of Grace' which were controlled by the Bishops. The issue of divine anthropomorphism was much debated by the Church fathers during the mediaeval period. In terms of official Church theology, Christian teaching followed the ancient Jewish notion that God could not be represented in human form, but at a popular level there was always strong pressure towards a physical human representation of the figure of Christ. Furthermore the emphasis on Christ's human characteristics was part of a Christian humanism which influenced both popular and official teaching (Le Goff, 1988).

It is often argued in sociological research that Western cultures have gone through an all embracing process of secularization and disenchantment, whereby the authority and dominance of the sacred realm has been contained and then eroded by urbanization, the growth of an individualistic culture, the spread of secular rationalist knowledge and by social diversity

associated with large-scale population migrations. This process of rationalization has been closely associated with the sociology of religion of Max
Weber, who, in *The Protestant Ethic and the Spirit of Capitalism* (Weber,
1930), identified religious asceticism as a major cultural source of modern
rationalism and secularism.

The Cartesian perspective on the individual as an isolated rational human
being was clearly associated with the growth of a Protestant world-view
within which the isolated individual is separated from God through the negation of the efficacy of ritualistic systems. Such a religious actor was forced to
discover God through a cognitive or rational understanding of the New
Testament hence this system of religiosity placed greater emphasis on literacy.
Within this historical framework, we can understand Descartes's vision of the
body as a machine as part of the more general movement within religious cultures which stressed the rational capacity of human beings to grasp the world
and to understand it through nonreligious means. This secular world was to
be controlled simply by the use of a neutral technology. Descartes therefore
drew a significant distinction between the soul and the body, regarding the
body as simply a machine directed by instructions from the soul.

Of course, Descartes's attempt to reduce medicine to mechanics and to
treat medicine as a form of pure physics ultimately collapsed, because it is difficult to explain what creates the body as a machine, what directs and how the
human machine can be distinguished from, for example, a combustion engine.
While these philosophical problems in Descartes's theory are now relatively
well known, for example, in the work of Georges Canguilhem in his *A Vital
Rationalist* (Canguilhem, 1994), Cartesianism as a view of the world became
a basic principle of Protestant individualism, scientific rationalism and
Protestant spirit which sought to dominate external nature through the application of instrumental rationalism. Cartesianism had a number of primary
characteristics such as dualism, reductionism and positivism. It is ironic that
contemporary interpretations of the philosophy of Descartes, claim that his
own position was characterized in fact by a 'dualistic interactionism' (Wilson,
1978). It is clear from a reading of Descartes's *The Discourse on Method* that
he believed that there was in fact a significant interaction between the body
and mind and that disease was the outcome of any disturbance in this interaction. Descartes's dualistic interactionism eventually came to evolve in the
natural sciences into a unitary and positivistic perspective of materialism in
which the disciplines that attempt to develop an understanding of events in
nature and society, body and mind were both isolated and specialized.

While conventional sociology characteristically regards modern Western
society as secular, this is a unidimensional and simplistic understanding of
contemporary society, where the question of the body has continued to dominate much of contemporary thought about the nature of the individual,
social organization and the natural environment. Given our contemporary
anxieties about pollution and the environment, it is possible to suggest that
the Green movement and other forms of environmental politics have given
external nature a sacred status. In terms of the work of the theologian Paul

Tillich we can define 'sacred' as that which ultimately concerns us. Nature and the human body and the natural environment have become major issues of human concern. Within the environmentalist movement, nature has replace the suffering body of Christ, and human cruelty is seen in terms of the quest for individual rewards through a psychology which gives legitimacy to material greed. These narratives of the environmental crisis have many aspects of the Christian doctrine of suffering in the secular idea of the risk society (Beck, 1992), a contemporary theodicy in which the collective dilemmas of the planet are products of the uncontrolled, indeed insatiable, drive for modernization within a capitalist civilization.

In the context of analysing the body as the sacred object of the modern somatic society, it is important to be aware of the growing criticisms of Cartesianism in both popular culture and in the social sciences. The crisis of the modernist project can be understood in terms of the transition to a postindustrial or postfordist or indeed postmodern reality. These transformations have been accompanied by a growing criticism of the naive and simplistic assumptions of rationalist Cartesianism as the dominant ideology of ascetic capitalism.

As we have already seen, the principal features of the Cartesian myth were the separation of mind and body, the subordination of body to mind, and the associated dominance of cognitive rationalization. 'I think, therefore I am' became the crucial slogan of the Cartesian social world and this Cartesian ideology set in motion a whole process of rationalization in Western cultures. We can conceptualize Cartesianism as an ideology operating within three domains. First there is the arena of thought and rational inspection which denied or excluded the force of the irrational, the magical and the superstitious. For example, Cartesianism denied the efficacy of evil spirits and demons, thereby leaving the rational mind dominant in the field of consciousness. Secondly, there was the regulation of emotions, sexuality and the affective life through the regulation and discipline of the human body. Foucault has referred to the emergence of 'governmentality' and discipline in the seventeenth century in his *Discipline and Punish* and *The Birth of the Clinic*. The human body was subject to a new regime of discipline or governmentality within which for instance, dietary practices became an essential element of the ascetic control or government of body.

Finally, Cartesianism in the form of instrumental rationalism or means–end rationalism, was associated with the growth of colonialism within which other cultures were subordinated to the instrumental control of Western technology and civilization. In Shakespeare's *The Tempest*, which apparently was modelled upon a set of real events in the colonization of the Caribbean, we approach a perfect illustration of the struggle between the rationality of the enlightened West and the irrationality of primitive societies as seen in the figure of Caliban. While Prospero represents the rational order of Western society, Caliban is a figure who stands for irrational lust, magic and superstition against the innocent and beautiful figure of Miranda, who is finally rescued from the grotesque embrace of Caliban as a result of the sovereignty exercised by Prospero's rational mind and will. While the island is disturbed by the storm and by the irrational desires of Caliban and his followers, Prospero eventually through a form of enlightened despotism establishes governmentality over his subjects.

Christian theology has played an important role in shaping this Cartesian secular rationalism, which I believe was an essential ideological underpinning for the rise of ascetic capitalism. In the seventeenth-century therefore there was a certain cultural compatibility between Pauline theology in its Protestant manifestations and Cartesian secularism. The role of asceticism was to liberate the soul from its entrapment within the body, that is from the limitations and problems of human embodiment. Diet was a government of the body, the aim of which was to create and establish the freedom of the soul or the mind from the slavery of the bodily senses. These disciplinary practices in the Christian tradition existed to produce the rational self, the network that Foucault has called the technologies of the self.

Within this ascetic tradition (of the classical and Christian eras), the body was seen to be a threatening, difficult and dangerous phenomenon. It had to be adequately controlled and regulated by cultural processes because the body was seen to be a vehicle or conduit for the unruly, ungovernable and irrational passions, emotions and desires. The necessity to control the body especially its lactations, excretions and reproduction is an enduring element within this Western philosophical tradition of religion. This Christian critique of the body was however to some extent a continuation of certain attitudes and values within classical Greek society.

Foucault has argued in *The Use of Pleasure* (1987, p. 21) 'a certain association of sexual activity and evil, the norm of proprietive monogamy, the condemnation of relations with the same sex, the exultation of continence' were fundamental assumptions or themes within Greek civilization. These sexual norms were a consequence of a more profound commitment to the importance of self mastery which was seen to be a crucial technology for all moral subjects. Within Greek life, moral virtue was closely associated with the lives and values of free rational men who regulated the public sphere to the political exclusion of women, young men and slaves. Within this Greek tradition the relations between men and their wives were not focused on intimacy or sexual relations but rather on relations of authority in terms of the control and government of domestic spaces. Because sexual pleasure threatened this world of rational self mastery it had been regulated and relegated from the public sphere. Indeed these forms of political separation and moral exclusion involved ascetic practices which were the foundation of citizenship. Women, slaves and boys were socially and morally inferior to rational free men because they were objects rather than subjects of sexual encounters. There was a close association between rational speech within the public sphere and ascetic standards of beauty which treated women as inferior in terms of rational discourse and sexual desire. This pattern of relations formed the basis of homosexuality in classical Greek civilization as a norm of desirable practice between men.

With the growing political importance of Christianity as a cultural force within the West, there was a further elaboration of the ascetic attitude towards the body, since as a consequence of the Christian definition of the body as evil, the body came to be more closely associated with Man as a fallen or defective creature. Indeed the human body was transformed into the notion of flesh, which in turn associated the human body with animality. While flesh stood in the same sphere as sub-human animality, the soul became the carrier and symbol of all forms of spirituality and rationality. In the monastic tradition, the body was given therefore a darker meaning as flesh as the metaphor of mankind fallen from grace and alienated from God. This unruly passionate body required the discipline of diet, meditation and religious practice.

However, following the creation mythology in Genesis of woman as the betrayer of Man, the female sexuality in particular was associated with the dark side of human existence. Women came to be associated with moral and religious temptation, with the story of the Fall of Adam from Paradise. In this Christian tradition, Man therefore came to be associated with the spirit and with reason, while woman came to be associated with matter and passion. This picture is obviously complicated by the importance of Mary's spirituality and her role as the mother of God. Nevertheless the emphasis on Mary's immaculate conception indicates the importance in theology of separating Mary's spirituality from the routine, everyday occurrence of motherhood and reproduction. Mary was untouched by the corporeal features of birthing and motherhood, because her embodiment could not threaten the spirituality of Jesus.

Of course these dichotomies set up a number of important tensions in Christianity because Jesus had to be of woman born in order to be human and to suffer in this world alongside mankind whom he had come to redeem. The dominant theme in Christianity therefore remained the idea of woman as the weaker vessel, as the conduit of evil. In the seventeenth-century this Christian mythology is supremely expressed in the encounter for example between Satan and Eve in Milton's epic *Paradise Lost*. In the form of a classical orator, Satan gains control of Eve by manipulating her sense of lack following the prohibition on the eating of the apple (Kendrick, 1986, p. 213).

In Weber's sociology of religion there is the central thesis that rationalization and calculation depended significantly on the theology of the Protestant sects which came through a series of unintended consequences to lay the foundations for the emergence of asceticism, moral control and capitalist accumulation. Rationalization involved the secularization of culture, the erosion of superstition, the decline of magic, the intellectualization of everyday life through the control and imposition of scientific reasoning, the calculation and regulation of bodies in the political interests of greater control and more efficiency, and the control of everyday life through the development of micro-bureaucratic techniques and practices. These cultural changes were closely associated with the rise of the market economy in which the calculation of the values and commodities was independent of tradition and religious teaching.

We can obviously interpret this rationalization project as an ethic of world mastery involving the control of the self through the regulation of the flesh, the subordination of matter to the training and perception of the mind, the taming and exploitation of nature in order to bring about a domination of natural wilderness and social reality and finally the control and subordination of the external societal environment by the West through an economic and political project of colonization. This ethic of world mastery as a project of modernity involved the institutions of regulation of government and through this political regime a regulation of bodies via dietary practices and other appropriate medical controls. This conquest of nature by the ethic of world mastery produced a new empiricism and positivism in the form of Baconian experimental science and pre-eminently in the shape of Cartesian rationalism. This ethic of world mastery operated at three levels, namely at the level of the individual (in terms of a mind/body dualism), at the level of society (in terms of a dualism between rational government and unruly protest) and at the level of inter-societal relations (in terms of the political dualism between friend and enemy).

The emergence of this modern ethic of control corresponded to the emergence of a philosophical project whereby the external world could be understood by rational enquiry and it is this project of rational enquiry which provides the historical link between Hobbes, Descartes, Locke and Kant. The expansion of European colonialism created especially at the political level the origins of a global society within which philosophical universalism could flourish. It was on the basis of this colonial world that the Enlightenment philosophers could with a sense of security and confidence, philosophically speculate about the essential and fundamental questions of truth, irrationality and beauty. Truth had become universal because the world had become a global environment. Indeed societies which diverged from these central notions of truth, reason and beauty were understood as deviations from a rational human culture, deviations which were associated with the themes of otherness and difference. Obviously colonialism and universal rationalism stood in an ambiguous relationship whereby enlightenment philosophy had so to speak to turn a blind eye to the exploitation which makes philosophy possible. There existed a complex relationship between the notion of subordinate and free peoples on the one hand and the subordination of the body to the mind on the other, a paradox or tension which lay behind the Enlightenment as a whole (Horkheimer and Adorno, 1973).

Against this world of political and social stability there were negative realms which had to be controlled by government or excluded by thought. Madness itself was opposed to reason and had to be ejected from society by the growth of the asylum, which Foucault explores in *Madness and Civilization* (1967). While mad men and women were physically excluded from normal society through the creation of prisons and asylums, irrationality and madness were excluded by Cartesianism from the world of philosophical reasonableness. If madness haunted the world of reason and spirituality, the world of nature was threatened by monsters, pathologies and other forms of deviation. The physical anomaly was a confusion in nature which threatened the order of things. It was indeed an affront to the Great Chain of Being.

These three issues (madness, illegitimacy and anomaly) came together in the figure of Woman, who in the medico moral legacy of Aristotle and Galen was characteristically regarded as a monstrous creature, indeed a secondary creation. Within this Western world therefore the medical code was typically combined with the Christian Creation mythology to provide a significant denigration of women as forces of disorder. Woman's sexuality therefore has to be understood as problematic within a patriarchal society, because it is necessary to guarantee the legitimate reproduction of men in order to secure the economic basis of society. In order to understand women's sexuality we need to comprehend the distribution of property within societies, regulated through the patriarchal structure of the domestic space and the nuclear household.

We can see therefore that patriarchy was a fundamental dimension of the modernizing process of which Cartesianism was a dominant ideology as we have seen, Cartesianism pre-eminently expressed the virtues and values of possessive individualism, of the rational enquiring mind and of mastery over the natural environment. If Cartesianism is the ideological expression of the modernizing project of which the body was a major target, then postmodernization has to be associated with fundamental changes in the structure of the household, the nature of capitalist production, changing relations with nature and the environment and with various processes of decolonization and restructuring of the global economy (Turner, 1990). Postmodernization expresses this disembedded characteristic of the body with regard to the household, property and reproduction. The postmodern body, so to speak, gives expression to this dismembered form of the body within a society dominated by consumerism, by a global economy and by the political struggles of a colonial world system. The postmodern emphasis on desire expresses a partial reversal of the traditional view of asceticism, control and production which is characteristic of early capitalism.

While Cartesianism may have been an appropriate ideological form in relation to the economic and social structure of early capitalism it would be quite mistaken to believe that this ideology of ascetic rationalism was never challenged or threatened by alternative social forces and belief systems. In Christianity, for example the relationship between women and spirituality was more complicated than I have so far suggested. Caroline Bynum's research on the tradition of female mystics in medieval society has been important in shaping the views of historians and sociologists with respect to the social role of women in the Church. In her *Fragmentation and Redemption* (Bynum, 1991) she demonstrates that any interpretation of female spirituality must be placed in a historical context where medieval theologians emphasized the idea of the continuity of the body (that is the absence of fragmentation) for the continuity of the self (namely the redemption of the soul). Physical resurrection was a fundamental assumption of mainstream medieval religious

thought, despite the critique of human existence as flesh which was fundamental to asceticism. Although woman was condemned in the Old Testament
mythology for her physical nature in the twelfth and thirteenth centuries
female mystics somatized their experiences of the divine. In the struggle
against heresy, the physical miracles of female mystics were employed by the
Church as a powerful counter argument to the doctrines of the heretics and
the ungodly. Religious ecstasy among women became a method of legitimiz-
ing the role of women in society. Indeed Mariology was a powerful
theological aspect of such arguments against heresy, and out of these conditions there developed the notion that both the Church (*ecclesia*) and
humanity (*humanitas*) were female in character.

### The Baroque and the Postmodern

If the Protestant ethic was indeed the spirit of capitalism (namely the cultural
form within which capitalist modernization was expressed), then we can
regard baroque culture as a form of oppositional struggle against the new patterns of possessive individualism and ascetic economic production. Many
features of the postmodern world were in fact present within the baroque culture of the seventeenth century which challenged the new ascetic Protestant
sects through a revitalization of traditional patterns of politics and social
relations (Turner, 1990; 1993). Baroque culture had been thrust into a global
arena by European imperialism which had a strong sense of the fragmented
and socially constructed nature of reality, which produced an articulate
notion of existential anxiety and the subjectivity of the self. Baroque art
employed parody and irony as rhetorical devices by which to express this
new sensibility.

Baroque culture may well therefore provide us with the archaeology of the
modern from within, because baroque culture, as the culture of the crisis of
seventeenth-century absolutism was opposed to the individualism of the
Protestant sects which writers like Weber, Mannheim and Parsons have identified with the origins of modern society and in particular with industrial
capitalism. The individualist religion of the seventeenth-century Protestant
merchants was counterposed to the centralized politics of absolutism with its
attachment to the counter-reformation spirituality. Liberal Protestantism was
identified through the works of Hobbes and Spinoza with the idea of individual rights, the social contract with the rational state and the legitimacy of
parliamentary government. Political absolutism supported the divine right of
Kings to rule arbitrarily. Thus the culture of the baroque period was a culture
of spectacle which attempted to mobilize the human senses to achieve a mass
commitment to absolutist monarchies (Maravall, 1986). Baroque culture was
a conservative culture which manipulated the masses through fantasy,
through elaborate music, through brilliant colour and sensual sculpture.

These social conflicts between religious individualism and political regulation were also expressed through the changing nature of baroque dance and

court life-style in the late sixteenth and early seventeenth centuries. Intellectuals like Montaigne were fascinated by the process of improvisation in art as a metaphor of political instability against a background of traditional political systems. These patterns of improvisation in dance culminated in burlesque ballet in the 1620s (Franko, 1993). For Montaigne, dance is an important illustration of the physical expressivity of the self. In late renaissance society, dance, where it took place between the sexes, presents society with an obvious problem of sexual liberty, if not licence. It also presented through improvisation the possibility of social and cultural diversity. For Protestants, dance was morally and socially reprehensible. The burlesque ballet (1624–7) opened up the possibility of cross dressing, feminization of male dancers and the representation of grotesque bodies. Baroque dancers expressed themes which were central to the baroque: melancholy, vanity and madness.

The luxurious sensuality of baroque public culture can be regarded, particularly in its public architecture and culture of display, as a mixture of high and low culture developing an early form of kitsch sensibility which was designed to trap the masses within a simulated culture. Thus the baroque culture of the seventeenth century can be regarded as a form of cultural industry which produced a mixture of high and low taste within a single theme. The primary images of the baroque period were centred on the ruin, the labyrinth, the library and various patterns of artifice. Visual deception and artificiality fascinated the baroque mind. The fascination with ruins and decay was an important aspect of the baroque sense of the artificial and the constructed character of reality but it was also associated with a deep sense of melancholy which pervaded this period. Montaigne's persistent concern with melancholy, the body and religious conduct was characteristic of this period (Starobinski, 1983). Robert Burton's *The Anatomy of Melancholy* of 1621 was also a basic expression of the baroque imagination, since melancholy ruins and decay all suggested or indicated a mentality of decadence and corruption for which the medical interventions of Burton and the medical notions of Montaigne were seen to be a cure.

The fact that the baroque mentality dwelt upon the images of Saturn, the classical ruin, the academic library and the garden labyrinths of the aristocracy was a consequence of this all-pervasive notion of political catastrophe and social collapse. It should not be surprising that the idea of catastrophe constituted an important aspect of the baroque. The philosophy of Gottfried Leibniz (1646–1716) has been closely identified with the growth of the baroque in this respect. Leibniz's doctrine of the monads and his view of theodicy have often been regarded as a cynical justification of absolutism as a type of arbitrary government. Leibniz's doctrine that we live in the best of all possible worlds can thus be seen as a legitimatization of centralized absolute power under the cloak of theodicy in which the universe is seen as a perfect labyrinth of 'windowless monads'.

The relationship between baroque culture and modern society can perhaps be seen through the work of Christine Buci-Glucksmann (1994) in terms of her notion of the theatricization of social reality. Just as baroque culture created the public spectacle as a system for subordinating the mass of the population to arbitrary control, so the modern world of consumerism can also be regarded as a form of spectacle. The sociological notion that modern reality is based upon spectacle and that the masses are seduced by the spectacular nature of modern popular culture was developed by the Situationnist International in France in the 1950s and 1960s. In 1977 Guy Debord published his *The Society of the Spectacle* which offered a general theory of the role of the spectacle in contemporary society. The Situationists argued that industrial capitalism was faced by imminent collapse through violent political struggle. These ideas were briefly made plausible by the student revolts of 1968 and many of their views of the importance of consumerism were taken up by postmodern philosophers such as Jean Baudrillard and J-F. Lyotard.

In his *In the Shadow of the Silent Majorities* (1983), Baudrillard has suggested that modern capitalism is impervious to social and political change and criticism because the mass acts as a black hole which absorbs all critical attack. The creation of a mass society by modern technologies of communication means that contemporary society is a kaleidoscope of whirling signs and symbols, a simulated world of self-referential signs which are divorced from any sense of social reality. In these postmodern theories of mass society the city is seen to be the fulcrum of modern social change. The baroque city provided the image of the endless labyrinth within which the new masses could hide and threaten authority by their very proximity. Throughout the nineteenth century, the city, particularly Paris and Berlin, generated the themes and images of urban social change and the focus of sociological enquiry. The artificial character of urban culture provided an important topic of sociology, particularly in the writing of Georg Simmel. In the development of postmodern theory, the artificiality and consumerist tendency of the modern city formed the basis of the work of Baudelaire and Walter Benjamin. The human body was fundamental to these urban consumerist images and spectacles. In this respect, we can regard mass sporting events such as the World Cup and the Australian Football League grand final match as baroque consumerism of the body.

## The Critique of Cartesian Authority

In the social sciences and the humanities in the late twentieth century the principal assumptions of modernism and modernization have been challenged by the growth of postmodern theory, which we can see as a critique more specifically of the Cartesian view of reality. Postmodernism embraces a radical scepticism towards the grand narratives or general assumptions of modernism by adopting a highly relativistic position on knowledge and belief, at least a relativism which challenges the modernist version of universalism. Postmodern epistemology has grasped a radical perspectivism with respect to the claims of rational knowledge and argues that there can be no single authority by which the world could be understood or apprehended.

Postmodernism has critically challenged the foundation of Cartesian ideology by its emphasis on the narrative quality of human knowledge and by its approach to the idea of social constructionism as a challenge to empiricist notions of reality.

The postmodern critique of Cartesianism has a number of philosophical bases. In particular much of the philosophical authority of Cartesian thought was questioned by the early rise of phenomenology of philosophers like Edmund Husserl who wished to defend the facticity of everyday life ('the despised doxa'), that is he rejected the subordination of everyday sensuality, facticity and practicality to the superiority of abstract cognition and rationalism. Cartesianism had rejected the emotionality of the human actor in favour of an instrumentalist rationalism which had the consequence of subordinating the habitus of everyday life to a superior cognitive order. Prior to twentieth-century phenomenology however, radical writers like Ludwig Feuerbach (1957) had, in his *The Essence of Christianity* in 1841, attempted to establish sensualism as a principle for attacking the rationalist legacy of German philosophy, particularly in the work of Hegel. As I argued in chapter 8 below, Feuerbachian sensualism was an important challenge to the legacy of rationalism and it was from Feuerbach that philosophers came eventually to contrast the Cartesian notion of 'I think, therefore I am' with an entirely new materialistic slogan, namely that 'man is what he eats'. Feuerbach produced a critique of the cognitive assumptions of Cartesian rationalism and he created a view of human beings as practical sensual beings who transformed themselves in the very process of reorganizing their inner environment of the senses and their outer environment of the natural world.

It was this Feuerbachian materialism which shaped the social thought of the young Karl Marx who came to develop his own materialistic version of Feuerbachian essentialism through an emphasis on the economic satisfaction of human needs. Marx therefore took the sensualist views of Feuerbach and converted them into a fundamental definition of human beings as creatures who labour on reality through the production and reproduction of their own reality. In the twentieth century the revival of this humanistic Marxism provided a new impetus to a sensualist approach to the human body and the human environment which challenged the university legacy of Cartesianism and the instrumentalist and economic view of Marx in favour of a collectivist Utopia of genuine socialism.

Although Marxism has been an important foundation for the anthropological critique of Cartesianism, the postmodern rejection of the Cartesian myth depends more heavily on the legacy of Nietzsche. Through his contrast between Dionysus and Apollo, Nietzsche developed his own view of the nature of human beings and their society in terms of an endless struggle between the principles of instrumental rationalism and the necessity for sensual satisfaction in the lives of human beings. Apollo stands for the principles of formalism, rationalism and consistency while Dionysus stands for the realities of ecstasy, fantasy, excess and sensuality. Within Nietzsche's paradigm, Protestant Christianity was seen to be the legacy of Apollonian rationalism and in terms of the Protestant sects there emerged an ascetic culture which regulated sexuality in favour of an ethic of hard work. Nietzsche argued that it was only through the reconciliation of these two dimensions that human beings could achieve any real balance in their lives, namely through a reconciliation of art and existence.

This view of the irreconcilable conflict between sensuality and rationalism was further elaborated in Freudian psychoanalysis where the subconscious regulation of the libidinous forces of human life gave rise to various types of neurosis. It was through this Freudian 'talking therapy' that the underworld of Dionysus could be turned into a positive force in society. Nietzsche's view of human beings has shaped much of twentieth-century social thought as a critique of the legacy of Cartesian rationalism (Stauth and Turner, 1988). Nietzsche's view of human beings now is regarded as an unfinished animal who requires cultural reality within which to work and live. This patriarchal image of man in the legacy of Nietzsche is evident in the work of philosophers like Arnold Gehlen in his influential *Man: His Nature and Place in the World* (1988). The philosophical anthropology which follows from this idea emphasized the world openness of human beings and their lack of instinctive specificity. In this philosophical perspective, human beings are homeless in the world because their genetic and instinctual structure do not fit them automatically or readily for a comfortable life on this planet.

Whereas the image of human beings in Cartesian mythology was one of control, domination and sovereignty, the image of the human being in this post-Cartesian mythology and postmodernism is the anxious, thus playful creature in search of personal satisfaction through new forms of intimacy and direct personal experience. Postmodern ethics in the world of writers like Zygmunt Bauman (1992) emphasize this frailty and vulnerability of human beings as lonely, homeless and alienated creatures, subject to the new world of consumerism.

We can see much of this intellectual growth in the last 150 years, a response to and typically rejection of this system of Hegelian philosophy. Certainly we can regard Kierkegaard, Schopenhauer and Nietzsche as representing in some sense existential rejection of Hegel's system. Nietzsche in particular was a philosopher against systems, and if we regard Kant and Hegel as inaugurating the modernist project in Western philosophy, postmodernism is a contemporary version of this rejection of Hegelian universalistic idealism. It is important to see Jürgen Habermas's defence of modernity and rationality as a contemporary defence of Hegel through a revised version of Marxism in the tradition of the Enlightenment philosophers. In *The Philosophical Discourse of Modernity*, Habermas (1987) indicates the validity of this interpretation when he argues that, while Horkheimer and Adorno in *Dialectic of Enlightenment* (1973) undertook a protracted struggle with Nietzsche, Heidegger and Bataille have assembled under Nietzsche's banner for the final struggle with the legacy of rationalism.

There is a general theme in Western social theory which posits a struggle or contradiction between nature and culture. We should not suggest that the expression of this theme was entirely coherent or that there were no variations on these themes. For example, it is significant that Foucault wanted to distance himself from this interpretation of the centrality of sexuality and he complained that he had been 'given the image of a melancholic historian of prohibitions and repressive power. . . . But my problem has always been on the side of another term: truth' (Foucault, 1988a, p. 111). For Foucault human sexuality is simply the truth of desire. Foucault did however want to show that in addition to the histories of economics and politics, 'it was also possible to write the history of feelings, behaviour and the body' (Foucault, 1988a, p. 112).

There is a similar way in which we can read Norbert Elias through a Freudian paradigm. Elias is not opposed to the civilization process, because he also regards civilization controls as beneficial to individual development (Elias, 1978). Although there are different traditions within philosophy, it is clear that Western thought has been profoundly shaped and influenced by a series of primary dichotomies namely between body/soul and nature/culture. This legacy can be defined in terms of a Cartesian ideology of ascetic control within urban capitalism. However, my argument in this introduction is that this dominance of Cartesianism with its separation of mind and body has been challenged directly by feminism, postmodernism and critical theory which are in turn philosophical and social consequences of major transformations in the nature of society, primarily towards the emergence of a postmodern or information society.

## The Project of the Self in the Late Modern Age

It has often been argued that possessive individualism with its focus on economic ownership was the dominant ideology of early capitalism. However there are now strong indications that this particular form of individualism has declined along with the changing nature of economic ownership and changes in the legal characteristics of individual property owners (Abercrombie, Hill and Turner, 1986). It does not follow, however, that our cultural world is no longer individualistic, but rather that in contemporary society traditional individualism takes a new form. We are still individualistic in that our orientation to reality assumes a knowing cognitive subject, but much of this cognitive rationalism has been replaced by an emphasis on emotionality, sensibility and sensuality. One can detect this emphasis in some recent sociological writing on the human body.

For example, Chris Shilling in his *The Body and Social Theory* (1993), Anthony Giddens in his *Modernity and Self Identity* (1991) and Anthony Synott in his *The Body Social* (1993) have argued that in contemporary societies the project of the self, as the principal legacy of individualism has now been converted into the project of the body. Shilling argues, for example, that with transformations in modern technology the human body has assumed a new plasticity whereby it can be readily transformed and recreated through surgical interventions. If the embodied self becomes the project of consciousness in modern societies, then the ageing of the human body becomes anathema to this social value of perpetual youthfulness. Ageing and death are denied and rejected within the new mythology of endless renewal, activity and youthfulness. Indeed, death becomes a threat to the stability of the system grounded on a view of the body beautiful, the body as pure fluidity and creativity.

This social view of the mobile self can of course be criticized because it raises problematic issues in the periodization of Western history and in the conceptualization of individualism. For example, the idea of the self as a personal project was implicit in much of the historical development of the Christian confessional and the Christian diary. The confessional enhanced the notion of self-consciousness and self-awareness through the ritualized dialogue of the confessional process (Hepworth and Turner, 1982). The idea of the self evolved further in Western society through the confessional literature associated with the rise of the romantic novel and finally developed through the rise of psychoanalytic practice. Although the historical framework of this argument about the emotional self and the body as a project can be criticized, the underlying thesis regarding the mobile self, selfhood as intimacy and the fluid body helps us to understand the impact of consumerism on contemporary subjectivity and the importance of sensuality for society as a whole.

This view of the sensitive self is somewhat different from the arguments presented by Foucault in his various commentaries on the body as a target or focus of disciplinary practices. Foucault developed the notion of biopolitics within which the individual was developed and cultivated through the emergence of a carceral society. In this framework it is the technologies of the self which produced consciousness through disciplinary arrangements. Although Foucault has provided a model of the self in modern culture through his historical analyses, the perspective of Foucault neglects the significance of consumerism, fashion and life-style on contemporary notions of selfhood. I have argued that the new self is far more mobile, uncertain and fragmentary than the bureaucratic image of the disciplined self in the work of Weber and Foucault because the modern self corresponds to and is produced by a new uncertainty, differentiation and fragmentation of the risk society (Beck, 1992).

In the postmodern cultural context within which the self evolves, the boundary of the self becomes uncertain and problematic and with technological change in the medical sciences, the body can indeed be restructured and refashioned to bring about profound changes of identity, including changes of gender. Thus, rather than thinking about the body as a regulated topic, we should conceptualize the body in a more fluid manner to allow for these important social changes in the wider social context.

In previous analyses, including *The Body and Society*, I have attempted to understand these general changes in social relations through a historical study of diet and dietary practices. In traditional Christian culture, diet was characteristically a method of regulating the self to protect the soul from the ravages of sexuality and thus diet was a governmental control or regimentation of the body. As such diet has a very close relationship to religion, which as a system of disciplinary practices binds the individual to the collective whole through such rituals as the Eucharist. These rituals which surrounded the body also defined the fabric of society in terms of a series of exclusionary and inclusionary practices. Religion and diet are a system through which the institutions of normative coercion operated, bringing about a regulation of individuals in the interests of social stability through a regulation of their bodies. Medical theory, practice and science played a major part in these moral views of the body and in shaping appropriate conduct towards the self and society.

One might have expected that with the secularization of Western cultures and the decline of religious institutions, views about diet will also have become more scientific and secular. One might assume that the ancient moral language of the medical diet would be transformed and replaced by more neutral and scientific language and nutrition. Of course there were important changes in scientific theories about nutrition in the nineteenth century and by 1840 German chemists had identified the three major chemical constituents of food, namely fats, carbohydrates and proteins, and they had begun to calculate the human requirements for each of these components. By the 1880s the scientific laws relating to the conservation of energy were being applied to all living organisms and nutritional scientists began to measure the potential energy in food and the energy which is expended in human labour in terms of calories, the units used in thermodynamics to measure mechanical work. It was typical in the late nineteenth century for scientists to regard food as a form of fuel for the human body conceptualized as a human engine.

Nutritional science became an important approach to assessing and improving the effectiveness of human labour. It was, for example, applied through rationing to institutionalized or captive populations in prisons or in the armed forces. Nutritional science became part of the moral campaign to improve the living conditions of the working class by reducing the cost of food by a more efficient means of cooking and eating. Behind the growth of nutritional science lay strong political and moral motives to regulate the working class (Crotty, 1995). It is interesting to reflect upon the ambiguities of the notion 'to ration', because a ration is an allowance of provisions, but it also means as a verb, to subject to a rationing scheme. It is also associated with the idea of rationalization, which in everyday terms means the development of planning and regulation. We can regard the growth of nutritional science as a rationalization of conduct, that is as an imposition of scientific norms and practices on everyday relations and activities. Within a broader context there has of course been a rationalization of the industry of eating and consumption through such developments as the McDonald's hamburger chain which has applied Henry Ford's production theories to the preparation of food to establish a standardization of patterns of consumption through the fast food industry. In fact the management principles involved in the

McDonaldization of the food industry can also be applied in universities, journalism and other areas of life resulting in a general McDonaldization of society (Ritzer, 1993).

While there has been therefore a scientific transformation of nutritional knowledge and practice, the social policy outcomes of nutritional science still contain an important moral dimension. For example, contemporary public health policy is heavily geared to promoting low fat diet, despite lingering scientific uncertainty about the effectiveness of such a strategy in reducing morbidity and mortality rates. A low fat diet public health program is based upon the idea that there exists, particularly in the male middle class an unhealthy stratum of individuals who are overweight and therefore at risk in terms of chronic illness and early death. These nutritional ideas clearly integrate well with current organizational theory about the importance of a good corporate image. The modern corporation has to be lean and mean and its corporate leaders are expected to have bodies which metaphorically indicate and embrace the ideology of organization fitness.

Of course, diet and self-regulation were for many centuries part of a religious discipline which aimed to control the soul. In modern societies asceticism is designed to produce an acceptable social self, particularly a self that conveys sexual symbolism. Looking good in our society means looking sexually attractive, and for women this has come to mean being thin. The modern soul or psyche is expressed through the body, that is through a sexually charged body image which is socially good. While nutritional scientists may have produced the slogan 'eat less fat' with the aim of reducing mortality rates in advanced industrial societies, 'eat less fat' is acceptable in popular culture because it is understood within a discourse of the sexual body. It is for this reason that the whole question of diet is inextricably bound up with the problem of modern personal identity where a good body-image is important for a good self-image. A youthful slim body is a major personal and social asset which must be maintained through the life cycle. Thus in contemporary society, this role of diet has been reversed because we diet in order to express our sensuality and our sexuality. The body as a project is incorporated within the world of fashion and consumerism so that we slim in order to look good in order to be attractive to others. In contemporary society the self is, as I have argued, a representational self, whose value and meaning is ascribed to the individual by the shape and image of their external body, or more precisely, through their body-image. The regulatory control of the body is now exercised through consumerism and the fashion industry rather than through religion.

These governmental, disciplinary institutions form a system of normative coercion, which returning to the scheme presented in Nietzsche's philosophy, we can regard the institutions of coercion as part of the Apollonian system of society which involves a rationalization and regulation of the self and social relations. By contrast the Dionysian cults of ancient society may be in functional terms replicated in contemporary popular culture where the liminality of individual and social relations is enhanced in postmodern culture by an emphasis on hedonism and sensuality. Thus the Apollonian regulation of bodies was clearly relevant to early capitalism with its culture of ascetic discipline and labour. In such a culture, work was given a primary value and leisure existed merely to offer some therapy for the reproduction of working bodies in the labour process. Thus in postindustrialism with postfordist economy, the intensive extractive industries are replaced by a new advertising, fashion and leisure industry where there is a new emphasis on labour mobility and occupational flexibility. With permanent unemployment and underemployment, the changing nature of the relationship between leisure and work is illustrated through the problematic status of retirement in contemporary societies. The body beautiful and youthfulness are emphasized in a society in which consumption and life-style are more prominent within a postindustrial economy and postmodern culture. These changes in late capitalism have produced a series of cultural contradictions in postindustrialism between hedonism in the arena of consumption and the remaining legacy of asceticism in the labour process. In the absence of clear employment opportunities and with the collapse of traditional work careers, young people generally, but particularly male working-class youth, have adopted rebellious life-styles and body-images which indicate that unemployability, punk life-styles and rock values signify this unavailability for work.

## Major Analytical Issues in the Sociology of the Body

This is a broad historical framework within which we can understand changing aspects of the body in contemporary society. However, the principal focus of this new introduction to *The Body and Society* is to consider a number of major theoretical problems in the contemporary development of the body in order to produce a greater consolidation of our sociological understanding, which in turn will lay the bases for a more structured and influential research agenda. In this section I will be concerned with various approaches to the body and in particular my intention is to address some of the underlying issues in the debate between social constructionism and foundationalism. In considering these analytical issues, I shall focus on two empirical issues, namely the question of left-handedness in human societies and the process of human ageing. There are many different traditions through which the body has been conceptualized which give an intellectual expression to the bewildering question of the body in society. While there are many traditions, we can usefully identify a limited range within which contemporary debate has evolved.

First there is the notion that the body is merely a set of social practices. This anthropological tradition expressed the notion that the human body has to be constantly and systematically produced, sustained and presented in everyday life and therefore the body is best regarded as a potentiality which is realized and actualized through a variety of socially regulated activities or practices. This idea of the body as social practice is fundamental to Erving

Goffman's study of such diverse phenomena as face-work, stigmatization and embarrassment in social life (Goffman, 1969). From the perspective of symbolic interactionism, Goffman showed that the body as a practice can betray one in public life by giving off information which is not controlled and which is potentially damaging or threatening to the social self.

This idea of the body as a set of social practices found its most coherent statement in the anthropological work of Marcel Mauss, who developed the concept of body practices to understand the nature of the self in its social context (Mauss, 1979). For Mauss the body is a physiological potentiality which is realized socially and collectively through a variety of shared body practices within which the individual is trained, disciplined and socialized. To provide an illustration which was adopted by Mauss himself, although the human body has the potential for walking, the particular form of walking which is produced within a given society or group is the outcome of training and practice. For example, there is a significant difference between the goose step and the waltz although both are based upon the potentialities of the human body for upright movement and flexibility. The work of eth-nomethodologists like Harold Garfinkel has also produced a distinctive view of the body as an organized set of practices which are relevant to sustaining social order at the micro level (Garfinkel, 1967).

These approaches to the body all assume some notion of the importance of everyday life practices in the production and maintenance of bodies and in that respect they can be seen as part of the legacy of Husserl's interest in the contrast between the everyday world and the rational world of the state and its basis in instrumental rationality and the sciences.

In contemporary sociology and anthropology this notion of the centrality of the body in everyday life practices has been expressed through the work of Pierre Bourdieu in his concern for the importance of the habitus in relation to social practice. In the work of Bourdieu on social class and the body we find particular attention given the fundamental differences in the nature of the body in varying everyday experiences, particularly of occupational groups in relation to their social habitus. For Bourdieu therefore the body has a certain cultural capital which is expressed through practices which are specif-ically directed at the outer body (Bourdieu, 1984). In his *The Body and Social Theory*, Shilling (1993) has attempted to build a theory of the sociology of the body on the basis of Bourdieu's notion of practice and habitus and Giddens's notion of structuration to create an approach to the sociology of the body organized around the idea of reflexive modernity. Although this approach is clearly attractive the idea of practice in both Bourdieu and Giddens can be criticized (Turner, 1992, pp. 67–98) for failing to reconcile a phenomenology of the body with a structuralist view of the causal impor-tance of class as structure. At least in his early work, Giddens's approach to action is rather like that of Talcott Parsons in that for Giddens the body in his *The Constitution of Society* (Giddens, 1984) is a condition of action rather than a feature of embodied actor. However from the point of view of socio-logical theory the idea that the body can be understood in terms of everyday practices is appealing and in *The Body and Society* I clearly adopted a similar approach, namely through Mauss's idea of potentiality.

One can adopt a foundationalist approach to the human body which avoids simplistic materialism and also allows us to understand how culture and social practices elaborate and construct the human body through endless social relations based upon reciprocity. In this respect we can avoid some of the dichotomies between action and innate behaviour and between culture and nature. Once more turning to the question of human movement the work of Oliver Sacks appears to express this idea very precisely: 'walking, at its most elementary, is a spinal reflex, but it is elaborated at higher and higher levels until, finally, we can recognise a man by the way he walks by *his* walk' (Sacks, 1981, p. 224). Because sociology is ultimately a social science of inter-action which pays especial attention to the question of the meaning of actions, the sociology of the body must be grounded in some notion of embodiment in the context of social interaction and reciprocity. An ade-quate sociology of action would thereby start with an assumption about the embodiment of the agent, and the role of this embodiment in the endless reciprocities of everyday life.

A second tradition in the sociology of the body conceptualizes the body as a system of signs, that is as the carrier or bearer of social meaning and sym-bolism. Although classical sociology neglected the issue of the body in social action, anthropology has by contrast always had a special concern with the body in relation to symbolism. Ritual preparation of the body, the scarifica-tion of the body and the cultural transformation of the body in rites of passage have been central topics of cultural anthropology throughout its existence as an academic endeavour. Anthropologists have clearly under-stood the importance of the body in conveying culture and shared meanings, especially in ritual, ceremony and religious practice. In modern anthropology this tradition of analysis has been elaborated and developed in the work of Mary Douglas (1966; 1970). In the work of Douglas, the human body is an important source of metaphors about the organization and disorganization of society. Thus disorganized bodies express social disorganization, as for example in magical attacks upon the body.

In her analysis of risk and taboo, Douglas has attempted to capture this idea of the danger to the external features of the body via social attack. Through this form of theorizing, Douglas has implicitly attempted to avoid a naturalistic conception of the body, since the body is merely a means of thinking about metaphorical relations. Thus, the dietary rules of Judaism, as they are expressed in Leviticus, are in fact means of thinking about social rela-tions and forms of expressing social harmony. The prohibition on pork was not therefore a hygienic measure of a public health policy but rather expressed an anxiety about the categorical confusions which were involved in the exis-tence of pigs which have cloven feet but unlike cows do not ruminate. In her more recent work the relationship between disaster, risk and pollution has been made more explicit, because pollution and risk express rather similar problems namely instabilities in modes of thinking and categorization

(Douglas and Wildavsky, 1982). In more recent feminist sociology, the idea of the female body as a metaphor of social relations is prominent. For example, Emily Martin (1989) has drawn attention to the interesting metaphorical significance of the idea of birthing as a form of labour in a capitalist society.

A third approach to the body interprets the human body as a system of signs which stand for and express relations of power. This approach to the body has become very prominent in feminist, medical and historical research. In particular Laqueur in his *Making Sex* (1990) has been concerned to show how in classical and mediaeval theories of the body there was a notion that there is one body but two genders, and that women are merely an inverted form of the male. Specifically the female genitalia were regarded as inverted forms of the male organ. Laqueur's historical research on this theory has been concerned to demonstrate how that conception eventually broke down, not as a result of scientific observation and experimentation but as the outcome of social and political change. In medical sociology, social constructionism has been concerned with the historical development of diseased entities, showing how again disease categories are a product of changing social relations of power and not of scientific advances. In my *Medical Power and Social Knowledge* (Turner, 1995) I studied a variety of disease categories and conditions such as anorexia nervosa and agoraphobia which in their historical development overtly expressed political conflicts between men and women in a context of changing capitalist relations. The social constructionist approach might be summarized by the slogan: 'the body has a history'.

Social constructionism has become the characteristic epistemological approach of feminist views of the body. For example, the perspective of Elizabeth Grosz in her *Volatile Bodies* (1994) demonstrates the influence of social constructionist views on the rejection of anatomy as destiny. Feminist writers have been concerned to reject the equation of the good body with the good person as the underlying principle of aesthetics in contemporary culture (Winkler and Cole, 1994). The underlying criticism is that the fashion industry and consumerism construct an ideal type of the female body which cannot be achieved by real women and that pornographic images of women in the fashion industry underpin and support the basic patriarchal power relations which continue to control men and women. Recent sociologies of masculinity have also questioned much of the underlying equation of power, sporting prowess and masculinity which have underpinned the mythology of male power in Western societies (Messmer and Sabo, 1990).

The theoretical impetus to criticize anatomical theories of gender difference have derived considerable inspiration from deconstructive techniques which were originally elaborated in literary studies and biblical criticism as means of questioning and criticizing textual authority. Deconstructivist techniques, anti-foundationalist epistemology and feminist theory have provided powerful tools for treating the body as a problematic text, that is as a fleshly discourse within which the power relations in society can be both interpreted and sustained. The critique of the text of the body therefore leads into a critique of power relations within society. These approaches have therefore drawn heavily upon deconstructivist philosophers such as Jacques Derrida and the psychoanalytic tradition of Jacques Lacan. These developments in contemporary social theory were particularly important in the work of Julia Kristeva (1982) and Luce Irigaray (1985). These approaches have provided a rich and influential stream of social theory (Crownfield, 1992; Jaggar and Bourdo, 1989; Moi, 1987).

These theories have been highly critical of phenomenological sociology, the social anthropology of the body and mainstream sociology of the body in so far as these social science approaches have assumed a foundationalist epistemology of the body. Radical deconstructionists are typically not concerned with the phenomenology of experience of sex, or the phenomenology of pain, or the social and individual experiences of illness, behaviour or the phenomenology of ageing. These approaches tend to be rejected or criticized as individualistic, psychological or subjectivist. The irony of much feminist critique of philosophy is that in identifying the body as a crucial topic of contemporary thought they deny the facticity and givenness of the body as a consequence of their deconstructionist approach.

The paradox is that the lived body drops from view as the text becomes the all-pervasive topic of research. An important illustration of this trend in radical deconstructivist politics can be taken from the work of Judith Butler, particularly her *Bodies that Matter* (1993). Butler employs the notion of interpellation from the Marxist philosopher Louis Althusser to understand gender as the creation of a subject through the act of hegemonic interpellation as the hailing of an individual as a subordinated person. Thus she argues that the 'reprimand does not merely repress or control the subject, but forms a crucial part of the juridical and social *formation* of the subject' (Butler, 1993, p. 123). In this approach therefore sex 'not only functions as a norm, but is part of a regulatory practice that produces the bodies it governs, that is, whose regulatory force is made clear as a kind of productive power, a power to produce – demarcate, circulate, differentiate – the bodies it controls' (Butler, 1993, p. 1).

In this theory the materiality of sex is the constant iteration and interpellation of sex within a hegemonic order of power which both produces and regulates bodies in social space. In this respect materiality is in fact power in its formative or constituting role as the organizing principle of the domain of sexuality. It is clear from this approach, that Butler's work suffers both from the positive and negative features of Althusserian structuralism. In particular it is not concerned with sex as potentiality, as material practice in a phenomenological sense or with sexuality as interpersonal experience. Theory as a political critique attempts to radically criticize orders of power as constituting subordinate subjects. Given these structuralist presuppositions, it is difficult to see how oppositional social movements might criticize or destabilize such hegemonic regimes and it has very little to say if anything about the experiential and affective dimensions of social practice and social relations. As with other forms of structionalism, it equates, to use the language of sociology, the role with the person thereby equating identity with the phenomenology of the person. This is an old sociological question of course, namely is the person merely a collection of roles or is the person the organizing principle which integrates and orchestrates given social roles?

I have suggested elsewhere (Turner, 1992) that it is valuable to distinguish between epistemological questions, which are to do with what can be known, from ontological issues, which are concerned with what can exist. Sociological theories of the body also have to address these two issues. Deconstructionist approaches are primarily concerned with epistemology and often have little to say about ontology, that is, they question the taken-for-granted nature of the human body via a critique of knowledge but they are not particularly interested in, for example, the phenomenology of experience within the context questions about ontology. Obviously these two questions cannot be entirely separated but it makes sense to draw a distinction between critiques of knowledge and critiques of ontology. For example, one could have a radically sceptical view of knowledge about the body but also be committed to foundationalism, that is to the ontologically given nature of the human body.

However, there are some important problems with constructionism which are often neglected by radical critics who use a relativistic approach to knowledge in order to bring about certain political effects, such as a critique of patriarchal power. In my own work I have tried to avoid these dichotomies by suggesting that while the relativism of the sociology of knowledge and deconstructionism is a useful critical starting point for social enquiry it is not necessarily its conclusion. In any event most sociologists are likely to be weak constructionists. In the case of medical sociology constructionism becomes more interesting when one begins to enquire about differences in disease categories in terms of how they might be differently constituted. In short, accepting constructionism does not lead inevitably to anti-foundationalism, because some phenomena are more constructive than others.

Some of the difficulties with constructionism can only become clearer when we start looking at specific illustrations. The sociology of ageing is an interesting test for Butler's view of materiality as the reiterative interpellation of the subject. In my argument Butler confuses the status and role, on the one hand, with the social and individual processes on the other. Now age as a social category is clearly socially constructed and it makes sense to say that individuals are interpellated by these categories regardless of their individual differences. For example, in traditional societies there was the notion of a systematic series of stages of life often referred to as 'the seven ages of man'. Thus in traditional societies there were a series of roles and statuses which individuals occupied at precise and given moments in their life careers, careers which were closely monitored and regulated by a hegemonic process of interpellation. These age categories differ significantly across cultures and societies. In many traditional societies, for example, old age was a status in life which conferred respect, wisdom and responsibility. In the twentieth century, the social dominance of age and age sets has been challenged by various youth movements which criticized and undermined the notion that age conferred wisdom and respect (Wohl, 1979). We can argue that in this sense age is socially constructed, that age as a category has a social history, and that the significance of age groups and age sets varies across cultures.

A radical criticism of gerontological categories could argue therefore that age is socially constructed in such a way as to express dominant power relations and that the deconstruction of age is an important feature of social criticism. However, these arguments have little or nothing to do with the phenomenology of ageing as an individual and social process of bodily transformation. The notion that age is socially constructed does not require us to deny that ageing as a biological process involves a decline in physical ability, the emergence of grey hair, a decline in skin texture, an increase in the brittleness of the skeletal structure, and eventually a decline in mental ability. We can therefore without contradiction believe that dementia and Alzheimer's disease are medical categories which socially construct the troubles of ageing into definite medical definitions which allocate power to professional social groups and believe that ageing is a real process taking place in the organic foundations of human embodiment. To argue that ageing has a foundationalist dimension would furthermore not prevent us from also believing that the phenomenology of ageing is dependent upon such issues as social class position, cultural capital, gender identity and ethnicity.

The human body has definite and distinctive biological and physiological characteristics which are resistant to a deconstructionist epistemology. One prevalent and persistent characteristic of organic life, for example, is asymmetry. In fact biological asymmetry appears to be a fundamental principle of life as such (Bock and Marsh, 1991). This asymmetrical nature of existence is expressed in human life particularly through a difference between left- and right-handedness. We know clearly from anthropological work that left- and right-handedness is an underlying principle of symbolic classification which suggests of course that the distinction between left and right is socially constructed.

For example, there is some evidence that left and right was a basic principle for the organization of castes in South India from the eleventh century, the distinction between left and right was an important aspect of magical amulets in Spain and that left and rightness were associated with the role of the evil eye in North African cultures, and that handedness has been a fundamental aspect of religious imagination and soteriology (Brun, 1963). There is clearly a cultural preference for right-handedness and right-sidedness but the prevalence of right-handedness in human populations is both systematic and uniform across time and space. Much of the contemporary research on handedness in human populations has been brought together in S. Coren's *Lefthander* (1992). The historical evidence suggests that handedness has been a preferred categorization of excellence across all human cultures and that there is remarkably little variation in the size of the left-handed population, namely around nine per cent. There is some variation by age and Coren discovered that with ageing left-handedness declines. According to his data, whereas at the age of ten about fifteen per cent of the population is left-handed, beyond the age of eighty less than one per cent is left-handed. This decline in left-handedness could be explained either by cultural pressure to reduce the prevalence of left-handedness or by left-handed people dying younger. Coren's research indicates that people do not transfer from left- to right-handedness and therefore there are medical conditions which explain the decline of left-handed people in the population over time. For example, one explanation is the higher level of accidents amongst left-handed people in the populations of industrial societies.

The point of this illustration is that while handedness is clearly a classificatory principle and while the moral importance of symmetry is socially constructed, there are biological and physiological dimensions to asymmetry which cannot be explained or minimized by social constructionism as an epistemological position. The conclusion of this argument is that it is possible for us to embrace a sociology of knowledge approach to body systems as classificatory aspects of human culture and also believe that there are foundational aspects of human embodiment, such as asymmetry, which we do not need to deny in developing a sociology of the body. An emphasis on the importance of the concepts of practice and habitus permits us to understand both the phenomenology of experience and the socially constructed characteristics of body as metaphor.

## Theoretical Consolidation and Research Agenda

In the past decade there has been a rich growth of volumes which address the importance of the sociology of the body. This evolving interest in the body in the social sciences is indicated by such publications as John O'Neill's *Five Bodies* (1985) and *The Communicative Body* (1989), Francis Barker's *The Tremulous Private Body* (1984), David Armstrong's *Political Anatomy of the Body* (1983), Don Johnson's *Body* (1983), Emily Martin's *The Woman in the Body* (1989), and Mike Featherstone, Mike Hepworth and Bryan Turner's *The Body* (1991). Although there is now a rich tradition of theoretical writing on the body, the existing state of the sociology of the body is in many respects both underdeveloped and disappointing. I have indicated some of my anxiety about this underdeveloped status of the sociology of the body in my Preface to Pasi Falk's *The Consuming Body* (1994).

One can note that, for example, the sociology of the body has been developed in a limited number of research areas. The first is the representation of the body and the study of the body as a system of metaphors, a development which I have explored through various references to anthropology and art history. Secondly there is a major focus on the sociology of the body in terms of questions of gender, sex and sexuality. There is a strong tradition of feminist research on the body and in more recent years there has been a significant growth in the sociology of masculinity which has also established a significant enquiry into male bodies (Connell, 1987; 1995; Cornwall and Lindisfarne, 1994). Finally, there is also considerable research into the sociology of the body in recent debates in the social sciences with respect to medical issues. My own contribution to the sociology of the body was in fact to provide a theoretical underpinning for medical sociology, on the assumption that medical sociology had become an applied empirical area of sociology divorced from mainstream sociological enquiry into the nature of social action. Although there are these important areas of research therefore, the mainstream business of the social sciences has not been as yet penetrated by an interest in questions relating to embodiment, body and bodily practices.

In addition to this, so to speak, lopsided development of the sociology of the body, there is a general anxiety that the sociology of the body has been confined to theoretical speculation and elaboration without creating a strong research tradition or research agenda. Both Bob Connell (1990) and Loic Wacquant (1995) have complained of an excessive devotion to theory in the absence of genuine research. Wacquant's research on professional boxers is therefore particularly important in developing a research basis to an understanding of how bodies are shaped and managed through training and exercise. Wacquant shows how Bourdieu's work on habitus and practice can be used as a powerful research framework for understanding the shaping and managing of bodies through a sporting regime. We still need a strong research agenda for developing a coherent sociology of the body.

Thus in the first volume of the new journal *Body and Society*, we attempted to outline a preliminary research programme which would identify a coherent set of topics and issues for the development of better sociological understanding of the embodiment of the social actor. This research programme included the symbolic significance of the body, the understanding of the active role of the body in social life and further explorations of the differentiation between gender and sex. These topics are, of course, well known issues in contemporary sociology. In addition, we argued that the relationship between the body and technology will be an important aspect of this evolving area of sociological enquiry. For example, by directly altering the infrastructure of the body, genetic engineering will be part of a radical set of techniques bringing about the condition of 'posthumans'. Furthermore, by constructing artificial devices, technology can contribute to an increase in our empowerment. Mechanical devices become part of the outer skin of the human body facilitating human control over the environment and leading to recent innovations in body–machine fusions or cyborgs. We also argued that the sociology of the body would continue to contribute to the development of medical sociology and the sociology of health and illness where a better understanding of the nature of embodiment would bring about a more sophisticated sociology of illness, conditions and disease categories.

Finally, there is the area of the sociology of sport where an understanding of the body is a crucial feature of sociological enquiry. Again the work of Bourdieu has been important in establishing the legitimacy of the sociology of sport as an area of sociological enquiry where embodiment is a critical topic (Bourdieu, 1993). In this research agenda, attention is drawn to the distinction between the objective and subjective body, that is between the subjective experience of embodiment and the physicality of the body. This distinction is an important component for the development of empirical research. Here again the sociology of ageing can be an important focus for the development of this tension between our subjective experience of embodiment and the objective processes of ageing and decline. This approach to ageing however must take into account the changing structure of the life process and the life career in a postmodern context, where the notion of the seven ages of man has broken down against a background of labour market flexibility, early retirement, increased leisure and structural unemployment. The subjective experience of ageing in the context of a postmodern life-style has given rise to important research on the 'mask of ageing' (Featherstone and Hepworth, 1991). The new gerontology of ageing processes in a postmodern society provides the research context for exploring new patterns of identity, intimacy and sensitivity and draws attention to the impact of consumerism and postmodern life-styles on the phenomenology of ageing.

Although the growth of the sociology of the body requires a strong empirical research agenda there are major theoretical and conceptual questions which have not been resolved. For example, in *The Body and Society* in 1984 I adopted an approach to the sociology of the body which drew heavily on structuralist notions in the work of Foucault to suggest four societal tasks which were important in the regulation of the body. This has been criticized by Arthur Frank (1991) and Nick Crossley (1995) for adopting an approach which is focused on what is done to the body rather than on what the body does. Crossley suggests that a 'carnal sociology', based upon the work of Merleau-Ponty, would overcome the dualism implicit in much of the sociology of the body by examining the impact of the body on social relations. Arthur Frank presents a similar approach to the body in adopting the work of George H. Mead to claim that the body becomes most aware of itself in the process of encountering resistance from others. As an alternative to my societal tasks scheme, he develops four dimensions concerned with control, desire, relations to others and self-relatedness. These criticisms are a valuable corrective to the underlying structuralism of my early work which I believe can now be corrected by a greater focus on the phenomenology of experience, as outlined in my *Regulating Bodies* (1992).

These critical debates in the sociology of the body suggest a series of steps which are necessary for developing a more adequate theoretical treatment of the body. First it needs a sophisticated understanding of the philosophical notion of embodiment which would be a method of exploring the systematic contradictions and ambiguities of the body as corporeality, sensibility and objectivity. The sociological notion of the body must embrace the idea of phenomenological experience of embodiment and the facticity of our place in the world. I have suggested that the tensions between a subjective and objective understanding of the body can be an important incentive to research in the sociology of ageing.

The question of embodiment is also fundamental to much of the sociology of the sport, as we have seen in Wacquant's study of boxing regimes. Secondly an embodied notion of the social actor and a comprehensive view of how the body image functions in social space is a necessary step in a genuinely socio-logical appreciation of how body images contribute to the occupation of social space and interaction with others. Thirdly, my approach to the sociol-ogy of the body is based upon an elementary understanding that sociology has to be based on an adequate appreciation of the reciprocity of social bodies over time and in space, that is an understanding of the communal and collective nature of embodiment. We are, to use a philosophical expression, always and already embodied. We do not have to develop a sociological appreciation of the physicality of the body since the 'natural body' is always and already injected with cultural understanding and social history. Finally the sociology of the body has to have a thoroughly historical sense of the body and its cultural formation, and much of this work has already been undertaken by historical studies of the medical regimes which have orga-nized and produced gender and sexuality. This research agenda and these theoretical topics will produce a renaissance of sociological research and enquiry in the postmodern period, when the body is indeed the principal topic of civilization in a somatic regime.

## References

Abercrombie, N., Hill, S. and Turner, B.S. (1986) *Sovereign Individuals of Capitalism*, London.

Armstrong, D. (1983) *Political Anatomy of the Body: Medical Knowledge in Britain in the Twentieth Century*, Cambridge.

Barker, F. (1984) *The Tremulous Private Body: Essay on Subjection*, London and New York.

Baudrillard, J. (1983) *In the Shadow of the Silent Majorities*, New York.

Bauman, Z. (1992) *Intimations of Postmodernity*, London.

Beck, U. (1992) *Risk Society: Towards New Modernity*, London.

Bell, D. (1976) *The Cultural Contradictions of Capitalism*, New York.

Bock, G.R. and Marsh, J. (eds) (1991) *Biological Asymmetry and Handedness*, Chicester.

Bourdieu, P. (1984) *Distinction: A Social Critique of the Judgement of Taste.* London.

Bourdieu, P. (1993) *Sociology in Question*, London.

Brun, J. (1963) *Le Main et l'esprit*, Paris.

Buci-Glucksmann, C. (1994) *Baroque Reason: The Aesthetics of Modernity*, London.

Burchell, G., Gordon, C. and Miller, P. (1991) *The Foucault Effect: Studies in Governmentality*, London.

Butler, J. (1993) *Bodies that Matter: On the Discursive Limits of Sex*, London.

Bynum, C.W. (1991) *Fragmentation and Redemption: Essays on Gender and the Human Body and Medieval Religion*, New York.

Campbell, C. (1987) *The Romantic Ethic and the Spirit of Modern Consumerism*, Oxford.

Canguilhem, G. (1994) *A Vital Rationalist: Selected Writings from George Canguilhem*, New York.

Connell, R.W. (1987) *Gender and Power: Society the Person and Sexual Politics*, Cambridge.

Connell, R.W. (1990) 'I am man: the body and some contradictions of hegemonic masculinity', in M.A. Messmer and D.F. Sabo (eds), *Sport, Men and the Gender Order*, Champaign.

Connell, R.W. (1995) *Masculinities*, Cambridge.

Coren, S. (1992) *Lefthander: Everything you Need to Know about Left-Handedness*, London.

Cornwall, A. and Lindisfarne, N. (eds) (1994) *Dislocating Masculinity: Comparative Ethnographies*, London and New York.

Crossley, N. (1995) 'Merleau-Ponty: the illusive body and carnal sociology', *Body and Society*, 1(1), pp. 43–64.

Crotty, P. (1995) *Good Nutrition? Fact and Fashion in Dietary Advice*, St Leonards.

Crownfield, D. (ed.) (1992) *Body/Text in Julia Kristeva: Religion, Women and Psychoanalysis*, Albany.

Debord, G. (1977) *Society of a Spectacle*, Detroit.

Douglas, M. (1966) *Purity and Danger: An Analysis of Concepts of Pollution and Taboo*, Harmondsworth.

Douglas, M. (1970) *Natural Symbols: Explorations in Cosmology*, London.

Douglas, M. and Wildavsky, M. (1982) *Risk and Culture*, Berkeley, California.

Durkheim, E. and Mauss, M. (1963) *Primitive Classification*, Chicago.

Elias, N. (1978) *The Civilizing Process*, Oxford.

Falk, P. (1994) *The Consuming Body*, London.

Featherstone, M. and Hepworth, M. (1991) 'The mask of ageing and the postmodern life course', in M. Featherstone, M. Hepworth and B.S. Turner (eds), *The Body: Social Process and Cultural Theory*, London, pp. 371–89.

Featherstone, M., Hepworth, M. and Turner, B.S. (eds) (1991) *The Body: Social Process and Cultural Theory*, London.

Feuerbach, L. (1957) *The Essence of Christianity*, New York.

Foucault, M. (1967) *Madness and Civilization: A History of Insanity in the Age of Reason*, London.

Foucault, M. (1987) *The Use of Pleasure. The History of Sexuality*, vol. 2., Harmondsworth.

Foucault, M. (1988a) *Politics, Philosophy, Culture: Interviews and Other Writings 1977–1984*, London.

Foucault, M. (1988b) *The Care of the Self. The History of Sexuality*, vol. 3., Harmondsworth.

Frank, A.W. (1991) 'For a sociology of the body: an analytical review' in M. Featherstone, M. Hepworth and B.S. Turner (eds), *The Body: Social Process and Cultural Theory*, London, pp. 36–102.

Franko, M. (1993) *Dance as Text: Ideologies of the Baroque Body*, Cambridge.

Garfinkel, H. (1967) *Studies in Ethnomethodology*, Englewood Cliffs, NJ.

Gehlen, A. (1988) *Man, His Nature and Place in the World*, New York.

Giddens, A. (1984) *The Constitution of Society*, Cambridge.

Giddens, A. (1991) *Modernity and Self Identity: Self and Society in the Late Modern Age*, Cambridge.

Giddens, A. (1992) *The Transformation of Intimacy: Sexuality, Love and Eroticism in Modern Societies*, Cambridge.

Goffman, E. (1969) *The Presentation of Self in Everyday Life*, London.

Green, B. (1993) *Gerontology and the Construction of Old Age: A Study in Discourse Analysis*, New York.

Grosz, E. (1994) *Volatile Bodies: Towards Corporeal Feminism*, Bloomington, Indiana.

Habermas, J. (1987) *The Philosophical Discourse of Modernity*, Cambridge.

Haraway, D.J. (1985) 'A manifesto for cyborgs', *Socialist Review*, 80, pp. 65–107.

Haraway, D.J., (1991) *Simians, Cyborgs and Women: The Reinvention of Nature*, London.

Hepworth, M. and Turner, B.S. (1982) *Confession: Studies in Deviance and Religion*, London.

Hertz, R. (1960) *Death and the Right Hand*, London.

Horkheimer, M. and Adorno, T. (1973) *Dialectic of Enlightenment*, London.

Irigaray, L. (1985) *This Sex Which Is Not One*, Ithaca.

Jaggar, A.M. and Bourdo, S.R. (eds) (1989) *Gender/Body/Knowledge: Feminist Reconstructions of Being and Knowing*, New Brunswick.

Johnson, D. (1983) *Body*, Boston.

Kendrick, C. (1986) *Milton: Study in Ideology and Form*, London.

Kristeva, J. (1982) *Powers of Horror: An Essay on Abjection*, New York.

Laqueur, T. (1990) *Making Sex: Body and Gender from the Greeks to Freud*, Cambridge, Mass.

Le Goff, J. (1988) *Medieval Civilization 400–1500*, Oxford.

Lewins, F. (1995) *Transsexualism in Society: A Sociology of Male to Female Transsexuals*, Melbourne.

Maffesoli, M. (1995) *The Time of the Tribes*, London.

Maravall, J. (1986) *Culture of Baroque: Analysis of a Historical Structure*, London.

Martin, E. (1989) *The Woman in the Body: A Cultural Analysis of Reproduction*, Milton Keynes.

Mauss, M. (1979) *Sociology and Psychology: Essays by Marcel Mauss*, London.

Messmer, M.A. and Sabo, D.F. (eds) (1990) *Sport, Men and the Gender Order*, Champaign.

Moi, T. (1987) *French Feminist Thought*, Oxford.

O'Neill, J. (1985) *Five Bodies: The Human Shape of Modern Society*, Ithaca and London.

O'Neill, J. (1989) *The Communicative Body*, Evanston, Ill.

Ritzer, G. (1993) *The McDonaldization of Society*, London.

Sacks, O. (1981) *Migraine: Evolution of the Common Disorder*, London.

Schilder, P. (1964) *The Image and Appearance of the Human Body*, New York.

Seidman, S. (1994) *Contested Knowledge: Social Theory in the Postmodern Era*, Oxford.

Shilling, C. (1993) *The Body and Social Theory*, London.

Starobinski, J. (1983) 'The body's moment', *Yale French Studies*, no. 64, pp. 273–305.

Stauth, G. and Turner, B.S. (1988) *Nietzsche's Dance: Resentment, Reciprocity and Resistance in Social Life*, Oxford.

Synott, A. (1993) *The Body Social: Symbolism, Self and Society*, London.

Turner, B.S. (ed.) (1990) *Theories of Modernity and Postmodernity*, London.

Turner, B.S. (1991) 'Recent developments in the theory of the body', in M. Featherstone, M. Hepworth and B.S. Turner (eds), *The Body: Social Process and Cultural Theory*, London, pp. 1–35.

Turner, B.S. (1992) *Regulating Bodies: Essays in Medical Sociology*, London.

Turner, B.S. (1993) 'Baudrillard for sociologists', in C. Rojek and B.S. Turner (eds), *Forget Baudrillard?*, London, pp. 70–89.

Turner, B.S. (1994) 'Preface', in P. Falk, *The Consuming Body*, London, pp. vii–xvii.

Turner, B.S. (1995) *Medical Power and Social Knowledge*, 2nd edn, London.

Turner, B.S. (ed.) (1996) *The Blackwell Companion to Social Theory*, Oxford.

Wacquant, L.J.D. (1995) 'Pugs at work: bodily capital and bodily labour among professional boxers', *Body and Society*, 1(1), pp. 65–93.

Weber, M. (1930) *The Protestant Ethic and the Spirit of Capitalism*, London.

Wilson, M.D. (1978) *Descartes*, London.

Winkler, M.G. and Cole, L.B. (eds) (1994) *The Good Body: Asceticism in Contemporary Culture*, New Haven and London.

Wohl, R. (1979) *The Generation of 1914*, Cambridge, Mass.

# INTRODUCTION TO THE
# FIRST EDITION

# BODY PARADOXES

There is an obvious and prominent fact about human beings: they have bodies and they are bodies. More lucidly, human beings are embodied, just as they are enselved. Our everyday life is dominated by the details of our corporeal existence, involving us in a constant labour of eating, washing, grooming, dressing and sleeping. To neglect this regimen or government of the body is to invite premature decay, disease and disorder. On a wider social scale, as Karl Marx constantly reminded us, society could not exist without the constant and regular reproduction of our bodies and without their allocation to social places. Over-production of bodies with respect to the availability of land and food resources, however, brings with it a different form of disorder: famine, war and pestilence. Although these observations are crassly obvious, few social theorists have taken the embodiment of persons seriously. Any reference to the corporeal nature of human existence raises in the mind of the sociologist the spectre of social Darwinism, biological reductionism or sociobiology. This study is based on the premise that these theoretical traditions are indeed analytical cul-de-sacs, which offer nothing to the development of a genuine sociology of the body. Although this sociological hostility to biologism is perfectly warranted, it does result in a somewhat ethereal conceptualization of our being-in-the-world. The human person is thus euphemistically referred to as a 'social actor' or 'social agent' whose character is defined in terms of their social location, their beliefs and their values. In this study, I want to argue that any comprehensive sociology must be grounded in a recognition of the embodiment of social actors and of their multiplicity as populations.

Sociological theory or, more broadly social thought, has been constituted around a number of persistent debates: what is the nature of social order, how is society possible, how is social control and regulation achieved, and what is the nature of the individual in relation to society? Within these constitutive debates, the body has made merely a cryptic appearance (Polhemus, 1978). In Social Darwinism and the functionalism of Talcott Parsons, the body enters social theory as 'the biological organism'; in Marxism, the presence of the body is signified by 'need' and 'nature'; in symbolic interactionism, the body appears as the presentational self; in Freudianism, human embodiment is rendered as a field of energy in the form of desire. The social sciences are littered with discourses on 'drives', 'needs' and 'instincts' which ooze out of the id. In this respect, much of sociology is still essentially Cartesian in implicitly accepting a rigid mind/body dichotomy in a period where contemporary philosophy has largely abandoned the distinction as invalid.

One objective of this study has been to reconsider much traditional sociological thought by incorporating the body within conventional debates about social order, social control and the stratification of societies. Having discussed the peculiar absence of the body in social theory, the theoretical core of my argument is then laid out in chapter 4, on 'Bodily Order'. Briefly, the thesis is that the classical Hobbesian problem of order can be re-stated as the problem of the government of the body. Every society is confronted by four tasks: the reproduction of populations in time, the regulation of bodies in space, the restraint of the 'interior' body through disciplines, and the representation of the 'exterior' body in social space. It will be evident that this schema is partly an attempt to provide a systematic version of the thought of Michel Foucault in *The History of Sexuality* (1981). Although this study of the body was crucially provoked by the debate over Foucault, my own argument tends to deny the originality of Foucault's contribution by, for example, bringing out certain continuities between Max Weber's notion of 'rationalization' and Foucault's discussion of 'disciplines' (Turner, 1982a). This chapter is fundamentally a reflection on the parallels between the idea of a regimen or government of the body and the regime of a given society. In order to illustrate this notion, a variety of 'disorders', especially in social subordinates, are taken as cultural indications of the problem of control. In particular the disorders of women – hysteria, anorexia and agoraphobia – are considered as disorders of society. The thrust of this claim is that any sociology of the body involves a discussion of social control and any discussion of social control must consider the control of women's bodies by men under a system of patriarchy.

In pre-modern societies, the regulation of the body has been closely bound up with the control of female sexuality in the interests of household authority and the distribution of property under a system of primogeniture. In the two chapters on patriarchy, the origins of patriarchy as an institution and as an ideology are traced through a discussion of the distinction between nature and culture, especially in terms of Christian theological attitudes to 'the flesh'. The argument is that the character of female subordination cannot be separated from the historical emergence of the patriarchal household and that female dependency cannot be separated from the government of the household in which junior men, women and children were under the authority of patriarchs. This debate is extended by a discussion of the classical form of patriarchy in the seventeenth century as it was expressed in the opposing views of Robert Filmer and John Locke. The main contention of chapter 6 is that the notion of patriarchal power cannot be uncoupled from the existence of the patriarchal household and that the development of capitalist society, by destroying the traditional household, undermines traditional patriarchy. In contemporary societies, we need a new conceptualization of the position of women and to grasp this departure from traditional patriarchy I propose the concept of 'patrism' as a defensive regulation of women.

On 'The Disciplines' of the body (chapter 7), the argument returns to certain themes raised by the recent work of Foucault, which is primarily concerned with knowledge and surveillance of bodies. Foucault's approach to the disciplines which regulate bodies can be perfectly illustrated by an analysis of the development of dietary techniques in Western societies. This history of diet attempts to show that dietary management emerged out of a theology of the flesh, developed through a moralistic medicine and finally established itself as a science of the efficient body. The principal change is that diet was originally aimed at a control of desire, whereas under modern forms of consumerism diet exists to promote and preserve desire. This conversion involved a process of secularization of bodily management in which the internal management of desire by diet was transferred to an external presentation of the body through scientific gymnastics and cosmetics. These social changes of the nature of the body cannot be analysed outside the context of changes in the production and distribution of commodities under a system of mass consumption. The sociology of the body thus leads into a discussion of new forms of the self in the debate on modern narcissism and new forms of regulation under mass consumption. As a consequence of these developments in contemporary capitalism, the conventional association between capitalist accumulation and the ascetic practices of the body becomes increasingly irrelevant as asceticism is replaced by calculating hedonism.

There is a common focus to these final chapters which is the nature of illness and disease as a test of the relationship between body and person. At one level, the argument is simply that we can never regard an illness as a state of affairs which is dissociated from human agency, cultural interpretation and moral evaluation. In everyday expression, the notion of 'having an illness' suggests an exterior state of affairs over which the victim has little control, because the malady is 'natural'. However, the notion of 'being ill' suggests a condition which is more immediate and proximate. This notion begins to explore the paradoxical dimension of illness as something over which we exercise choice, but which we cannot control in terms of its direction and consequences. In order to develop a framework for this analysis of paradox, I consider many of the contradictory features of anorexia nervosa in a society where thinness is a norm of aesthetical value. In the course of an analysis of human agency in 'women's complaints', it is important to consider some aspects of Ludwig Feuerbach's contribution to Marxism, since it was Feuerbach who first made popular the aphorism 'Man is what he eats'. These chapters develop this debate about the status of disease as disorder by an examination of the traditional distinction between 'disease' and 'illness'. It is argued that this dichotomy presupposes the validity of a distinction between 'nature' (in which disease is located) and 'culture' (in which sickness occurs). Since the 'nature/culture' dichotomy is relative and unstable, it is argued that we cannot regard 'disease' as a neutral, technical amoral category. What will count in a society as 'sin', 'crime' or 'disease' depends on a variety of conditions such as the intellectual division of labour and the separation of areas of knowledge by professional specialization. The emergence of 'disease' as a special category is the product of professional dominance and secularization.

The central problem of this book is ultimately the question 'what is the body?' and this question cannot be answered without an enquiry into social ontology. As a conclusion to this enquiry into embodiment, I outline two related but distinctive approaches to ontology from Marx and from Nietzsche. In Marx, universal human nature is defined in terms of the fact that men, in the generic sense, labour collectively on nature to satisfy their needs and in the process transform themselves into sensuous, practical, conscious agents. Nature exists as an independent reality, but it is constantly transformed and appropriated by human labour with the result that 'nature' too becomes a social product. Marx avoids a relativist position by suggesting that what is universal to the human species is the need to satisfy their needs through the appropriation of nature and what we all share in common is the transformative potential of *praxis*. Although Nietzsche also believed that knowledge exists in the interests of practical needs, what we know does not exist independently of language; language is our first and our last appropriation of reality. Since knowledge is variable by virtue of the variable grammar of language, what exists is also language-dependent. Our being is the product of classification – a position which is related to Bishop Berkeley's immaterialism (existence is perception) and a position which forms the basis of Foucault's argument that the body is the product of classificatory knowledge and of power. These separate ontologies produce two contrasted accounts of the body. In Marxism, the body is both the vehicle and the site of labour; it exists but it is constantly transformed by human agency. In Nietzsche, our corporeal existence does not pre-date our classificatory systems of knowledge and thus the body is nothing more and nothing less than a social construct. The idea that the body is a construct has radically profound implications for the debate about gender and sexuality, but it is not necessarily a pessimistic conclusion since what has been constructed can also be deconstructed.

The germ of this book arose out of two publications on dietary management (Turner, 1982a, 1982b) and a speculative article on the relationship between medicine and religion (Turner, 1980). In the course of writing this study, a variety of publications on the body have recently emerged, partly in the wake of Foucault's emphasis on the significance of the bio-politics of the body. Of this recent eruption of studies, the most pertinent are *The Political Anatomy of the Body* (Armstrong, 1983), *Bodies of Knowledge* (Hudson, 1982) and *The Civilized Body* (Freund, 1982). This recent sprinkling of books on the human body does not constitute a theoretical movement and therefore there is still a strong justification for arguing that the body is absent in social theory, especially in sociological theory. The point of this study is not to provide a decisive or definitive account of the body, but to indicate what a general theory of the body would have to take into account and to note those areas of sociology – medical sociology, the study of patriarchy, the nature of social ontology, the sociology of religion, the analysis of consumer culture and the nature of social control – where the question of the human body is especially prominent.

The study of the body is of genuine sociological interest and it is unfortunate that much of the field is already cluttered by trivial or irrelevant intrusions – neo-Darwinism, sociobiology, biologism. By contrast, there have been important developments in phenomenology, anthropology and existential philosophy which converge on the notion of human embodiment and which provide a basis for a sociology of social being. The theoretical influences on this particular study of the body are numerous, but those approaches which are primarily concerned with the meaning of the body and its social construction and manifestation have been the most formative – Oliver W. Sacks in *Awakenings* (1976), Norbert Elias in *The Civilizing Process* (1978), Mikhail Bakhtin in *Rabelais and his World* (1968), Susan Sontag in *Illness as Metaphor* (1978) and Georg Groddeck in *The Meaning of Illness* (1977). Although in anthropology there is a considerable tradition of research on body rituals, I have not drawn extensively on anthropological materials or approaches, since my main focus has been on the body in urban, secular, capitalist society. The main exception to this exclusion of anthropological perspectives has been the seminar paper by Marcel Mauss on 'Body techniques' which was first published in 1935 in the *Journal de Psychologie Normale et Pathologique* and recently edited in *Sociology and Psychology* (Mauss, 1979). We cannot, however, raise the question of the meaning of the body without, as sociologists, attempting to locate those meanings within a broader framework of social structure and historical change. In pursuit of that framework, I have been forced to reconsider Parsons's interpretation of the materialism of Thomas Hobbes, Marx's interpretation of the sensualism of Ludwig Feuerbach, Feuerbach's appropriation of Jakob Moleschott's theory of nutrition, the impact of Nietzsche's physiologism on Foucault's philosophy and the importation of Rousseau's naturalist romanticism into the work of Claude Lévi-Strauss, Richard Sennett and French structuralism.

The sociology of the body also opens up fascinating digressions and side-paths, which are temptingly bizarre. Of these labyrinths, I shall mention Bishop Berkeley's fascination with the body, despite his alleged immaterialism, and his commitment to the therapeutic virtues of 'tar-water', which was probably the ultimate proof of his aphoristic thesis '*esse* is *percipi*'. He even wrote poems about it:

Hail vulgar juice of never-fading pine!
Cheap as thou art, thy virtues are divine.

Fortunately this digression has already been brilliantly followed in John Wisdom's *The Unconscious Origin of Berkeley's Philosophy* (1953). There are some digressions one ought to follow but which are outside my competence: these are the histories of the nude, cosmetics, pornography, gymnastics, architecture and fashion. At least some aspects of these cultural developments are being discussed in the journals *Theory, Culture & Society, Salmagundi* and *Telos.* The body is absent in theory, but everywhere in embodiment.

In attempting to write about the body, it is impossible to avoid its contradictory character and I have attempted to express these contradictory features by a variety of paradoxes. We have bodies, but we are also, in a specific sense, bodies; our embodiment is a necessary requirement of our social identification so that it would be ludicrous to say 'I have arrived and I have brought my body with me.' Despite the sovereignty we exercise over our bodies, we often experience embodiment as alienation, as when we have cancer or gout. Our bodies are an environment which can become anarchic, regardless of our subjective experience of our government of the body. The importance of embodiment for our sense of the self is threatened by disease but also by social stigmatization; we are forced to do facework and body-repair. Our bodies are a natural environment, while also being socially constituted; the disappearance of this environment is also my disappearance. Furthermore, it is not simply a question of the singular body, but the multiplicity of bodies and their social regulation and reproduction. Bodies both in the singular and in the plural are locations, as Mary Douglas (1970) has lucidly demonstrated, of profound feelings and categories of purity and danger. Bodily secretions cannot be easily ignored and, for that matter, they cannot be easily controlled; this is one source of those persistent metaphors relating to the body politic and the human body.

In writing this study of the body, I have become increasingly less sure of what the body is. The paradoxes illustrate the confusion. The body is a material organism, but also a metaphor; it is the trunk apart from head and limbs, but also the person (as in 'anybody' and 'somebody'). The body may also be an aggregate of bodies, often with legal personality as in 'corporation' or in 'the mystical body of Christ'. Such aggregate bodies may be regarded as legal fictions or as social facts which exist independently of the 'real' bodies which happen to constitute them. There are also immaterial bodies which are possessed by ghosts, spirits, demons and angels. In some cultures, such immaterial bodies may have major social roles and have important social locations within the system of stratification. There are also persons with two bodies, such as mediaeval kings who occupied simultaneously their human body and their sovereign body. Then there are heavenly bodies, the geometry of bodies in space, the harmony of spheres and corpuscular light. Given this elusive quality of 'the body', it is perhaps appropriate that the Old English *bodig*, corresponding to the Old High German *botah,* is of unknown origin. Like *ontic, bodig* is everywhere and nowhere. The body is our most immediate and omnipresent experience of reality and its solidity, but it may also be subjectively elusive. As one migraine victim has painfully observed:

> I have the feeling that my body – that bodies are unstable, that they may come apart and lose parts of themselves – an eye, a limb, amputation – that something vital has disappeared, but disappeared without trace, that it has disappeared along with the place it once occupied. The horrible feeling is of nothingness nowhere. (Sacks, 1981, p. 90)

The body is at once the most solid, the most elusive, illusory, concrete, metaphorical, ever present and ever distant thing – a site, an instrument, an environment, a singularity and a multiplicity.

The body is the most proximate and immediate feature of my social self, a necessary feature of my social location and of my personal enselfment and at the same time an aspect of my personal alienation in the natural environment. One feature of my argument is that our attitudes to the body are at least in part a reflection of the whole Christian tradition in the West. My body is flesh – it is the location of corrupting appetite, of sinful desire and of private irrationality. It is the negation of the true self, but also an instructive site of moral purpose and intention. Its health is also my moral well-being, since salvation involves two activities, namely to salve the body and to save the soul. Although we are familiar with the notion of 'the flesh', in Western culture it would be unusual to refer to the body as 'meat'. The body is enveloped by restrictions and taboos; it is flesh but not to be eaten and it is meat but not to be cooked. Flesh can be consumed ritually in the body and blood of the Lord Jesus Christ, whose flesh is the salvation of my flesh. The procreative activity of my parents is the model of His reincarnation of the flesh, but much of this imagery has been dissolved in the capitalist society. What we confess is the contradictions of the mind rather than the appetitive and vegetative nature of the flesh. Since, according to Montaigne, we live in a world of quotations from quotations, it is difficult to escape the paradox of Dionysus and Apollo. Those critics of modern consumption like Herbert Marcuse often sound like the ghost of the Puritans: the pleasures we enjoy are always false because they represent power over the body.

The final paradox is our sexuality – our attempts to organize it, control it and express it. It is, therefore, odd that sociologists have largely ignored sex. There are scattered texts – Barthes's *A Lover's Discourse* (1982), Carter's *The Sadeian Woman* (1979), or Brown's *Love's Body* (1966) – but there is no solid tradition which we could call 'the sociology of sex' and, as a general rule, the major sociologists have studiously ignored the issue. This gap is probably one further reason for taking Foucault seriously. The rationale for this particular book is, as a consequence, relatively modest; it attempts to describe the absence of a sociology of the body and it suggests a variety of approaches, especially through the analysis of disease and desire, by which the body could become a focus of sociological enquiry.

# 1

# The Mode of Desire

## Needs and Desires

Human beings are often thought to have needs because they have bodies. Our basic needs are thus typically seen as physical: the need to eat, sleep and drink is a basic feature of people or organic systems. It is also common in social philosophy to recognize needs which are not overtly physical, for example the need for companionship or self-respect. 'Need' implies 'necessity' for the failure to satisfy needs results in impairment, malfunction and displeasure. The satisfaction of a need produces pleasure as a release from the tension of an unresolved need. The result is that 'need' is an explanatory concept in a theory of motivation which argues that behaviour is produced by the search for pleasure and the avoidance of pain. In Greek philosophy, the Cyrenaics and Epicureans placed great emphasis on the satisfaction of pleasures as a criterion of the good life. In utilitarianism, the notion of the hedonistic calculus became the basis of Bentham's political philosophy: the good society is one which maximizes the greatest happiness of the greatest number. The problem is that not all pleasures appear to be necessary and many of them appear to be destructive and anti-social. Human capacity for pleasures appears infinite, including self-flagellation, homosexual rape, torture, plunder and pillage. The philosophical solution has been to distinguish between good and bad pleasures, between real and false needs. For example, the outcome of the debate about pleasure and virtue in Greek philosophy was that 'we should try to live a frugal life in which necessary desires are satisfied, and natural but not necessary desires given some place, while vain desires are outlawed. Such a life would naturally be virtuous' (Huby, 1969, p. 67). While a person may gain sadistic pleasure from the pain of others, these pleasure-giving activities are not regarded as conducive to a good society based on companionship and these pleasures are thus regarded as vain and unnatural. There are at least two problems with this position. The first is that I am an authority on my own pleasures and therefore individuals may not be easily persuaded that their private pleasures are somehow false. Secondly, the argument equates 'desire' with 'need'.

Although the analysis of desire has a long history in philosophy (Potts, 1980) and although 'desire' is often associated with 'appetite', it is important to be clear that a theory of desire is not the same as a theory of need. For example, Freud's psychoanalysis was primarily a theory of desire and cannot be translated into a Marxist anthropology which is essentially a theory of need. The difference is that need implies an object which satisfies the need, the object of the need being external to it; desire cannot be finally satisfied since desire is its own object. The view of desire provides the basis of Freudian pessimism, because desire cannot be satisfied within society. The Oedipus myth signals this impossibility for desire. The satisfaction of needs can be the criterion of the good society, whereas the satisfaction of desire cannot. *Concupiscentia* and *ira* are thus corrosive of that friendship which the Greeks saw as the cement of social groups as well as the basis of individual virtue.

**Wisdom and Friendship**

Sociology is literally the wisdom or knowledge (*logos*) of friendship (*socius*). The task of sociology is to analyse the processes which bind and unbind social groups, and to comprehend the location of the individual within the network of social regulations which tie the individual to the social world. While sociology is a relatively new addition to the social sciences, the notion that friendship is the ultimate social cement of large-scale social collectivities, like the state, is relatively ancient. In *The Symposium* Plato gave full expression to the Greek ideal of friendship as that social condition which overcomes the anti-social desires for personal possessions and competitive eminence. The aim of the individual and the state should be the cultivation of virtue and happiness rather than the satisfaction of desires which are the springs of disharmony and envy. The order (*kosmos*) within the individual is necessary to the ordering (*kosmios*) within the large social world and both are intimately connected to friendship. It was Eros which was the force capable of bridging the gap between the two essential elements of reality – rationality embodied in Apollo and irrationality in Dionysus (Jaeger, 1944). The interior of the individual reflects the anatomy of society as a contest between desires, of which envy is especially prominent, and reason (Gouldner, 1967). Both Eros and friendship are necessary to fuse these disruptive and corrosive features of the psyche and society. We can see then that the roots of Western philosophy lie in two related issues: the struggle between desire and reason, the opposition between the binding of friendship and the unbinding pressures of individuation.

There is much that separates Plato's philosophical enquiry into the nature of friendship and the sociological analysis of social bonding, but, as I shall show, there is also much continuity. More importantly, the world in which Plato existed has been transformed by two events which are crucial to this particular study: Christianity and the industrial revolution. Given the strong chiliastic dimension of early Christianity, the primitive church posed a sharp and decisive opposition between the world and the spirit. The cultivation of the body could have no place within a religious movement which was initially strongly oriented towards the things of the next world. Early Christianity may have inherited from gnostic Essenism the view that creation was corrupt and worthy of moral condemnation (Allegro, 1979). After the destruction of Jerusalem and the absence of the messianic Return, the Christian church was forced to accommodate to the existence of Roman imperialism, but it retained what Weber called inner-worldly asceticism, that is a strong hostility to the things of this world. To some extent the emphasis in Pauline theology on the sinfulness of sex was reinforced by the adoption of Aristotelian philosophy which was similarly hostile to women.

Within the Christian ascetic tradition, sexuality came to be seen as largely incompatible with religious practice. In particular, sexual enjoyment is a particular threat to any attempt to create a systematic religious response to sinfulness. This problem of subordinating sexuality to a rational life-style forms the basis of much of Weber's view of the origins of religious intellectualism and rationalization. The argument is that 'ascetic alertness, self-control, and methodical planning of life are seriously threatened by the peculiar irrationality of the sexual act, which is ultimately and uniquely unsusceptible to rational organization' (Weber, 1966, p. 238). One 'solution' to this dilemma of human existence was the division of the religious community into an elite which withdrew from the world in order to abstain from sexuality and the mass which remained embedded in the profane world of everyday society. The laity reproduced itself within the restrictions of organized monogamy. The elite withdrew into celibacy and monasticism, recruiting its members through vocations rather than carnal reproduction. Sexuality, even within the limitations placed upon family life by religious norms, was thus a lay activity, permitting monks and priests to follow a life of rational control over the flesh. As a result of this severity towards sexual sinfulness, the human body was transformed from the occasion for sin to its very cause. The body became the prison of the soul, the flesh became, in the words of Brother Giles, the pig that wallows in its own filth and the senses were the seven enemies of the mind (Black, 1902). To control the body, the ascetic movement in Christianity turned ever more rigidly towards rituals of restraint – fasting, celibacy, vegetarianism and the denial of earthly things.

**The Mode of Desire**

Corresponding to every economic mode of production, it is possible to conceive of a mode of desire. In *The Origin of the Family, Private Property and the State*, Engels (n.d.) argued that, within the materialist perspective of history, every society has to produce its means of existence and reproduce its own members. An order of sexuality thus corresponds to an order of property and production. The mode of desire is a set of social relations by which sexual desire is produced, regulated and distributed under a system of kinship, patriarchy and households. These relations of desire determine the eligibility of persons for procreative roles and legitimate sexual unions for the production of persons. The mode of production of desire consequently has social, political and ideological dimensions; for example, sexual ideology interpellates persons as sexual objects with appropriate relations for the consumption of sexuality (Therborn, 1980). It can be argued that the mode of production produces social classes as effects of property relations (Poulantzas, 1973). Similarly, a mode of desire specifies a classification of 'sex groups', of which gender is the principal dimension dividing the population into 'men' and 'women'. However, the dominant sexual classification also designates 'boys' as subordinates who are not eligible for reproductive functions – they may be, of course, appropriate objects of desire. In modern terminology, we can suggest as an initial starting point that every mode of production has a classificatory system of sexual desire – a discourse which designates appropriately sexed beings and organizes their relations. It is this social discourse which specifies eligible sexuality not the dictates of human physiology.

Marx (1974, vol. 1, pp. 85–6n) argued that in the feudal mode of production it was Catholicism which constituted the dominant ideology of feudal social formations. It is possible to re-express Marx's view by claiming that in a feudal mode of production there has to be an ideological regulation of sexuality corresponding to the specific economic character of feudal societies and that it was Catholic sexual discourse which provided the dominant mode of desire. Human agents live their sensual, sexual experience via the categories of a discourse of desire which is dominant in given societies, but this discourse of desire is ultimately determined by the economic requirements of the mode of production. The discourse has a grammar specifying who does what to whom and it is this grammar of sex which designates the objects and subjects of sexual practices. It is clear that this rendition of Marx is an attempt to bring together an Althusserian analysis of modes of production (Althusser and Balibar, 1970) and a Foucauldian outline of discursive formations (Foucault, 1972). This study of the body departs from these perspectives in two crucial features. The first is that both Althusser and Foucault have little to say in any detail of the resistance of either individuals or classes to forms of regulation and surveillance, despite frequent reference to resistance to discourse. Secondly, structuralist analysis of discourse either ignores the effectivity of discursive formations or takes their effects for granted. To show that a discourse is prevalent is not to show that it is wholly effective (Abercrombie, Hill and Turner, 1980).

Every society has to reproduce its population and regulate it in social space; at the level of the individual, sexuality has to be restrained and persons have to be represented. These four problems may have a different prominence and salience in different societies depending on the nature of the economic mode of production. In feudal societies, especially for the dominant landowning class, the reproduction of the dominant class depended crucially on the regulation and restraint of the sexuality of subordinate members of the household. The conservation of land depended on the stability of inheritance through legitimate male heirs; a discourse of desire was necessary to secure these economic objectives and this discourse was primarily patriarchal and repressive. These features of the discourse were contained predominantly within Catholic morality which aimed to repress pleasure in the interest of reproduction. This is not to suggest that mediaeval attitudes towards women were all of a piece; woman was both Eve (the cause of all our woe) and Mary (the source of spiritual power) (Bernardo, 1975), but the principal feature of the social position of women in feudal society was dependency and subordination within the household. In the seventeenth century, 'a roving woman causes words to be uttered' and this pronouncement applied to nuns as much as it did to married, noble women (Nicholson, 1978). A woman's place was next to the hearth with her master's progeny. This mediaeval discourse promoted legitimate sexuality and separated it from desire. Within this context, the confessional assumed especial importance (Hepworth and Turner, 1982); it was a ritual for the production of the truth of sex (Foucault, 1981), but to establish the truth of sexuality it had to understand the error of pleasure. Much can be learnt, therefore, about feudal sexual discourses by an analysis of the teaching of the penitentials on marital and extramarital coitus.

For mediaeval Christian theology, any act of coitus which did not result in the insemination of the woman was a 'sin against nature'. The sexual act was to be devoid of pleasure and therefore if a man enjoyed his wife the act was regarded as equivalent to fornication. These 'sins against nature' included not only sodomy, bestiality and masturbation, but also coitus interruptus. These were unnatural because they did not result in insemination and their primary motivation was pure pleasure. The same arguments applied to concubinage and extramarital sexuality, especially where these were undertaken with primitive contraceptive measures. The confessional manuals also proscribed certain sexual positions which increased pleasure and decreased the likelihood of conception. The condemnation of extramarital sex combined a variety of notions; it was associated with pleasure, with contraception and with unnatural positions. In addition, it implied that husbands would unwillingly become the parents of children whom they had not fathered. There was a danger therefore that property would pass to offspring who were not in reality legitimate. The order of legitimate sexuality would not correspond to the order of property relations.

It is very easy, as a result, to discover in these mediaeval texts a discourse of desire which separated pleasure from property. The sociological question is, however, to discover whether these discourses had real effects on social behaviour. Since it was impossible to form a household without sufficient capital, there are commonsense reasons for believing that young couples would adopt coitus interruptus for pleasure where procreation was economically precluded (Flandrin, 1975). Marriage was thus regarded as an economic and political contract between families for the conservation of a land-owning class; the marriage bed was devoid of pleasure. Since procreative activities were confined to these contractual unions in marriage, desire had to find its location elsewhere.

## Asceticism

In mediaeval times, the attempt to create a rational and systematic regimen of denial was largely confined to the religious orders who, as it were, practised asceticism on behalf of the lay man. Expressing this differentiation in spatial terms, reason was allocated to the internal domain of the monastery, while desire ran rampant in the profane world of the lay society. In this respect, we could perceive the principal argument of Weber's *The Protestant Ethic and the Spirit of Capitalism* (1965) as an account of how the Reformation took the ascetic denial of desire out of the monastic cell into the secular family. Protestantism thus sought to break the distinction between the elite and the mass by transforming elite practices into everyday routines of self-control. Abstinence, the control of passions, fasting and regularity were thus held up as ideal norms for the whole society, since salvation could no longer be achieved vicariously by the labours of monks. The disciplines and regulations of the family, school and factory thus have their historical roots in the redistribution of monastic practices within the wider society. The monastic cell was installed in the prison and the workshop, while ascetic practices spread ever outwards (Foucault, 1979, p. 238).

Of course, the attempt to impose monasticism as a general secular norm of restraint necessarily led to resistances. The history of English sexual culture can be seen as a pendulum swing between restraints on sexuality and relaxations in moral behaviour. The Puritan revolution of the seventeenth century was followed, with the Restoration, by a new liberalism in sexual conduct. The return to a more rigid sexual life-style in the late eighteenth and nineteenth centuries was followed by a new permissiveness which has been the dominant theme of contemporary society (L. Stone, 1979). To some extent these restraints on sexuality also corresponded to restraints on the table. From an ascetic point of view, eating and sexuality are both gross activities of the body. Eating, especially hot, spicy foods, stimulates sexual passion. To control sexuality, Protestants attempted to regulate the body through a regimen of dieting. The Puritan Revolution of the seventeenth century was thus also accompanied by a series of restraints on food, cuisine and consumption. Spices were banned and major festivals, such as Christmas, ceased to be occasions for secular enjoyment; the festivities surrounding Twelfth Night were also crushed. With the collapse of the Cromwellian era, the social revolt 'against Puritanism is shown in the excesses that took place at court. Often important banquets and entertainments were inclined to relapse into dissipated orgies, the honoured guests spattered in cream and other beverages' (Pullar, 1970, p. 128). While in the nineteenth century cookery became increasingly the object of domestic science, eating itself was still clothed with a certain Puritan prudery. Like sex, eating for nineteenth-century women was something more to be endured rather than stimulated.

Max Weber's sociology of Puritanism is normally interpreted as an argument about the ascetic origins of capitalism. In these introductory comments, it has been suggested that, more widely examined, Weber's analysis of Christian asceticism is in fact about the rationalization of desire. There were many dimensions to this process of controlling desires. Certain institutions were developed to subordinate internal passions to reasonable controls – monasticism, celibacy, monogamy, castration. Desires were regulated by routines – vegetarianism, dieting, exercise, fasting. The passionate side of human personality was subject to scientific enquiries, and technologies were developed to prevent various forms of 'self-abuse', especially in children. Human energy could be safely channelled through vocations in the world; the sexual conquest was directed towards economic triumphs in business and commerce. Festivities, festivals and carnivals which were historically occasions for orgiastic release were originally suppressed by Puritanism and then prohibited by the routines of industrial capitalism. Public and collective festivals were gradually replaced by more individualized and private pastimes. In Weber's sociology of rationalization, there is the argument that the whole of life becomes increasingly subject to scientific management, bureaucratic control, discipline and regulation.

There are, however, at least two problems with Weber's analysis of capitalism. Asceticism provided a suitable cultural norm for capitalists who had to deny themselves immediate consumption in the interests of further accumulation. The requirement of investment for future profits precludes full enjoyment of present wealth. For the worker, it is different. Because they are separated from the means of production, they are forced to labour to live under conditions of what Marx referred to as the 'dull compulsion' of their existence. The problem of capitalism as a system is, however, that there also has to be consumption of commodities otherwise the circuit of commodity capital becomes blocked and stagnates. With the growth of mass production, the rationalization of distribution in the department store and the post-war boom, capitalism also had to develop a consumption ethic, which in many ways is incompatible with the traditional norms of restraint and personal asceticism. Weber's account of capitalism ends with the arrival of early, competitive capitalism in which desire is still denied in the interests of accumulation. Late capitalism, by contrast, is organized more around calculating hedonistic choices, advertising, the stimulating of need and luxury consumption. Late capitalism does not so much suppress desire as express it, produce it and direct it towards increasing want satisfaction.

The second problem with Weber's account is that while early capitalism transferred the monastery into secular society, it also bifurcated the secular world into a private sphere of use-values and a public sphere of exchange-values. Desire was relegated to the world of the intimate, private citizen, while the public realm became increasingly dominated by the norms of rational calculation and instrumental knowledge. Such a division largely corresponded to the division between men and women, the latter being the vehicles of emotion, need and intimacy. There is, therefore, a social division of emotions which runs alongside the social division of labour, rendering women the custodians of the intimate and the private (Heller, 1982). This spatial division is, however, further disrupted by the increasing involvement of women in production, the transformation of the nuclear household and the decline of male-dominated, labour-intensive industries with the deindustrialization of late capitalism.

## Desire and Reason

The legacy of both Christianity and industrialization is the prominence of bipolar oppositions in thought and culture between the body and soul, the body and mind, matter and spirit, desire and reason. These classificatory oppositions are true not only of society, but of the basic forms of thought in Western culture and philosophy. It is not surprising that these distinctions should come to play a major part in sociological thought itself. Social thought has been modelled around the notion that human beings are simultaneously part of nature insofar as they have bodies and part of society insofar as they have minds. Social contract theories from Hobbes onwards resolved this dilemma by arguing that, as a rational animal, it was in the interests of men to form binding contracts in order to have security inside society. In forming contracts, men give up certain natural rights and submit to authority, whether in the person of the king or a government, to achieve some respite from the insecurities of their natural condition. The notion that civilized life requires certain basic restrictions and restraint has subsequently become a widespread theme of sociological and psychoanalytic thought. Freud, for example, treated the incest taboo, a prohibition on sexual intercourse between affines and a resulting guilt complex in the Oedipus complex, as the original basis of social grouping:

> The tendency on the part of civilization to restrict sexual life is no less clear than its other tendency to expand the cultural unit. Its first, totemic, phase already brings with it the prohibition against an incestuous choice of object, and this is perhaps the most drastic mutilation which man's erotic life has in all time experienced. . . . Fear of a revolt by the suppressed elements drives it to stricter precautionary measures. A high-water mark in such a development has been reached in our Western European civilization. (Freud, 1979, p. 41)

Since Freud's attempt to analyse taboo as the basis of civilized life, both psychoanalysis and anthropology have reconceptualized totemism as a system of classifications. Language not prohibitions constitutes the division between culture and nature, but the same theme of desire versus power is central to much recent structuralist analysis.

Language is an impersonal system of communication in which we surrender our individuality. Language represents the authority of society over the unconscious. Thus in the work of Lacan (1977) language is the basis of the alienation between the self and the world, and this alienation involves a division between the infinity of our desires, which are denied by social conventions, and the finitude of our demands which are allowed by society. Similarly in the work of Foucault, there is an opposition between power/knowledge which is localized in every authoritative institution and freedom/irrationality which is implicit in every deviant resistance. Madness had to be banished from the realm of reason by Descartes just as the mad have to be removed and confined in society. Fundamentally, the control of madness involves the control of passions. Foucault's quotation from

François Boissier de Sauvages's *Nosologie méthodique* of 1772 neatly restates the classic opposition between desire and order:

> The distraction of our mind is the result of our blind surrender to our desires, our incapacity to control or to moderate our passions. Whence these amorous frenzies, these antipathies, these depraved tastes, this melancholy which is caused by grief, these transports wrought in us by denial, these excesses in eating, in drinking, these indispositions, these corporeal vices which cause madness, the worst of all maladies. (Sauvages, 1772, vol. *vii*, p. 12, in Foucault, 1967, p. 85)

The imposition of reason over desire and the internment of the insane corresponds to a new apparatus of control in the asylum and a new horizon of knowledge in the sciences of man.


## Homo Duplex

Despite the trend in sociology to see all human attributes as the product of social determinism, sociology and social thought are often founded upon a concept of *homo duplex* in which the individual is a complex balance of asocial passions and social reason. For example, Durkheim, who is often regarded as the sociological determinist *par excellence*, also adhered to the model of double-man. *The Elementary Forms of the Religious Life* argued:

> Man is double. There are two beings in him: an individual being which has its foundation in the organism and the circle of whose activities is therefore strictly limited, and a social being which represents the highest reality in the intellectual and moral order that we can know by observation – I mean society. (Durkheim, 1961, p. 29)

The role of culture is to impose on the individual the collective representations of the group and to restrain passions by collective obligations and social involvements. Without cultural restraint, the individual is under certain circumstances driven by excessive expectations towards anomic suicide. The conservative dimension to both Durkheim and Freud is therefore the view that society is bought at the cost of sexuality. For Durkheim, that cost was both necessary and desirable. Since man is both a member of nature by virtue of being an organism and a member of society by virtue of culture, some solution has to be found to this Jekyll-and-Hyde duplexity. For Durkheim, the restraint and regulation of man-as-body was to be found in the coercive nature of moral facts.

Although many social theories presuppose a dichotomy between mind and matter, soul and body or reasons and passions, their account of and solution for that duplexity are highly variable. While there is a theoretical link between Durkheimian sociology and modern structuralism, there are also important differences. For example, Foucault does not see power as always constraining; indeed he regards power as productive and enabling. Power does not so much deny sexuality as produce it for purposes that lie outside the individual. One feature of modern society to which Foucault draws attention is the idea that every individual, as a crucial feature of their social identity, must have a single, true sex. One has to be either male or female, since the hermaphrodite is a false or pseudo sex. It could have been imagined that what really mattered was 'the reality of the body and the intensity of its pleasures' (Foucault, 1980b, p. vii), but changes in law, juridical status, medical science and the administrative apparatus of society in the period 1860 to 1870 began to force individuals to have unambiguous sexuality. Behind the medical enquiries of the period, there existed a moral project that suggested people with dual sexuality were capable of indecency. While in contemporary society it is accepted that one may change one's sex, the notion that finally everybody must be either male or female is not dispelled. In this sense, it can be said that medical knowledge and medical power produce sex as a category of necessary identity rather than denying or removing it.

There are important indications in modern philosophy, especially phenomenology, that the traditional dichotomy of mind and body is false and in need of rectification. Whereas Cartesian philosophy set up an opposition between the body as a machine and the mind as rational consciousness, we cannot properly regard the body as an unconscious thing, since the body 'is both an object for others and a subject for myself' (Merleau-Ponty, 1962, p. 167). I both am a body and have a body, that is an 'experienced body'. Alternatively, much radical thought regards the opposition of body and mind as an aspect of social power, which subordinates desire to reason for purposes of authoritarian control. For such writers, the liberation of society presupposes the emancipation of the body and its passions from both psychic and social control. There is thus a long tradition of critical thought which advocates sexual freedom as essentially a political act of opposition. The critique of patriarchal power, the harmful consequences of sexual asceticism, the liberating character of pleasure and the denunciation of the element of prostitution in marriage have been themes linking together a wide variety of writers – Charles Fourier, Havelock Ellis and Wilhelm Reich. Although their positions are theoretically diverse, they are linked together by a certain eccentricity and utopianism – Fourier's communes and Reich's organismic box are clear illustrations of both. Of modern writers, Herbert Marcuse has provided a more coherent account than most of the denial of pleasure which capitalism allegedly requires.

**Play and Pleasure**

Marcuse departed from much of the orthodox core of traditional Marxism by asserting that labour, far from being the source of all value, was simply burden. Play and pleasure have to be restrained and subordinated to guilt in capitalist society in order to prevent any 'irrational' diversions from the centrality of productive labour. Hence, for Marcuse, play and sexuality have a revolutionary potential which has been seriously neglected by critical theory (Geoghegan, 1981). Marcuse adopted the basic framework of Freudian psychology to explain the processes of social control in capitalism where the superego controls libidinous drives under the watchful eye of the state and the family. Capitalism, however, comes to depend on 'surplus repression' which goes far beyond what might be regarded as the necessary constraints on individuals as members of society as such (Marcuse, 1969). These moral and political restraints on human sexuality were being gradually undermined by economic changes in late capitalism – particularly automation – which made the traditional pattern of work and the family increasingly irrelevant to capitalist economic processes. Social freedom requires sexual freedom; both freedoms were being made possible by capitalist economic change. The main threat to these potentialities came from the commercialization and commodification of sex, which rendered sexuality profitable. In *Eros and Civilization* (1969), therefore, perverts replaced the proletariat as the principal agents of change within a capitalist society.

Marcuse's reinterpretation of Marx via Freud raises a problem which is central to all social theories grounded on an opposition between desire (as liberation) and reason (as restraint). There are two dimensions to this. The first was clearly expressed by MacIntyre (1970, p. 47): 'What will we actually *do* in this sexually liberated state?' The second relates to this, namely that sexually liberated men may find their desires satisfied via dominance and pornography at the expense of women. The liberation of desire is implicitly the liberation of male desire which fails to provide any explanation of the location of women in a society where men through economic changes are either driven out of work by structural unemployment or liberated from work by automation.

The naive argument in favour of sexual liberation cannot adequately cope with the problem that sex can be a commodity – prostitution and pornography – that reinforces rather than questions prevailing social relations. Pornography is, however, paradoxical in providing both the illustration of the commercialization of sexual relationships and also the critical reflection of power and dominance in sexuality. The commodification of sex lends support to the argument that modern society is a pornographic society, 'a society so hypocritically and repressively constructed that it must inevitably produce an effusion of pornography as both its logical expression and its subversive, demotic antidote' (Carter, 1979, p. 86). What most liberationist accounts of sexuality, such as Marcuse's, fail to confront is the problem of pornography as the expression of (male) desire over instrumental rationality. The fascination of de Sade and Sadism is that Sadism both expresses the inherent power conflict that trails alongside sexual freedom and by expressing power unmasks an over-romanticized view of male/female encounters:

> An increase of pornography on the market, within the purchasing capacity of the common man, and especially the beginning of a type of pornography modelled in that provided for the male consumer but directed at women, does not mean an increase in sexual licence, with the reappraisal of social mores such licence, if it is real, necessitates. . . . When pornography abandons its quality of existential solitude and moves out of the kitsch area of timeless, placeless fantasy and into the real world, then it loses its function of safety value. It begins to comment on real relations in the real world. (Carter, 1979, pp. 18–19)

The pornographic utopia then begins to act as the mirror-image critique of the 'natural' but exploitative relations between men and women within the domestic sphere of the home. Marcuse's approach to desire/reason with its emphasis on play as liberation fails to take adequate notice of pornography as a practice of power which, only under special circumstances, acts as a platform of criticism and change.

One interesting absence from the critique of instrumental reason and its subordination of desire in the tradition that links together Fourier and Marcuse is the absence of children (Bell, 1980). This absence is one very strong indicator of the fact that the conventional or traditional debate about reason/desire is a debate among men which submerges or obliterates the connection between desire and reproduction. Children are almost entirely absent from the sexual utopias of men. The liberation from restraint often appears therefore as a one-sided male liberation from surplus restraint on the id. In writing the history of desire we would in fact have to write two histories, male desire versus female desire. Both Marxist and critical theory have been peculiarly blind to the social division of desire in terms of gender and patriarchy. This study of the sociology of the body hinges as a result on the masculine control over female desires. The liberation of sexuality has to ground itself in an analysis of how desire versus reason has been institutionalized within a sexual division of labour which also involves a social division of emotions.

Foucault recognizes that in the modern period, a sexual identity is imposed on us and this sexual identity has to be either male or female – an issue which he explores in the story of *Herculine Barbin* (1980b). Yet in his major work on sexuality there is no significant attention given to gender divisions and how cultural divisions are elaborated onto the physiological difference between men and women. For Foucault, sexuality is a unity which can have one history; we do not talk about the history of sexualities, because in Foucault's account the body is implicitly the unified datum upon which knowledge and power have their play. This assumption is widespread in the literature. In *Bodies in Revolt*, Thomas Hanna, for example, while recognizing the difference between male and female sexual roles, can still refer naively to sexuality as 'a centrum for human experience' (1970, p. 287). Similarly, while Deleuze and Guattari (1977) have attempted to criticize the notion of desire as merely lack or absence, they also imply that desire (the id) is ultimately a unity.

## Capitalist Bodies

Our attitudes towards sexuality, women's social roles and gender are in part the arcane legacy of feudal Christianity and the requirements of property relations in modes of production based on private appropriation. Our attitudes have also been shaped by the ancient history of family life and patriarchal household. In late capitalism, these attitudes in many ways no longer conform to the actual requirements of the economy or to the social structure of a capitalist society which is organized around corporate owner-ship. Because property and investment are now concentrated in corporate bodies, family capitalism no longer plays a major role in industrial economies. Capitalism no longer requires the unity of the family in order to guarantee the distribution of property. Although capitalism may still require the house-hold as a unit of consumption, it is not a requirement of capitalism that these households should be of the nuclear variety. The ascetic mode of desire is thus not pertinent to contemporary forms of capital accumulation and largely inappropriate to individual consumption. The factory floor must have social regulations to insure continuous and efficient production, but even in the case of productive arrangements it is perfectly possible to de-skill the labour force and replace it with the dead labour of machinery. Modern capi-talism tends to foster hedonistic calculation and a narcissistic personality. Consumer culture requires not the suppression of desire, but its manufacture, extension and detail.

The theoretical and moral reaction to these new possibilities of mass pleasure has been varied and complex. One common position is to see that new culture as essentially an ideological incorporation of the working class into capitalism; the new consumerism is simply the old 'bread and circus' approach to domination. Much recent analysis of consumption is in this respect largely negative (Baudrillard, 1975; Lefebvre, 1971; Marcuse, 1964); modern consumption is seen to produce a passive, subordinated population which is no longer able to realize its 'real' needs. Despite its critical tone, the analysis of consumerism often assumes a conservative stance. The argu-ment that consumerism encourages narcissism can implicitly embrace a nostalgic adherence to the family, the work-ethic and patriarchal authority (Barrett and McIntosh, 1982). The critique of modern leisure and con-sumption can also be puritanical, neglecting the element of personal freedom which some modern technology makes possible (Kellner, 1983). The critique of consumerism is thus a version of the dominant ideology thesis (Abercrombie, Hill and Turner, 1980) in which consumers are uniformly incorporated by all commodities. It is simply not the case that consumers inevitably absorb the meaning and purpose of mass advertisements (Ewen and Ewen, 1982).

There is, of course, another version of the incorporation argument. The hegemonic control of capitalism over desires and needs is exhibited in the sit-uation that capitalism can survive and tolerate individual deviance and social pluralism; the tolerance of capitalism is oppressive. However, if capitalism can survive successfully in a context of widespread sexual permissiveness and personal freedom in the market place of commodities, then it is reasonable to conclude that capitalism does not require massive ideological supports. It operates through political, economic and legal regulation of the population. The paradox of the hegemonic argument is that capitalism enjoys the hege-mony of permissiveness which it does not actually require. In my view, the argument can be expressed in a more cogent sociological form: capitalism no

longer requires hegemony in sexual and personal domains, and this is precisely why cultural pluralism is characteristic of late capitalist societies. What
capitalism does achieve is the commodification of fantasies and pleasures.
There has been a rationalization of desire through the supermarket, advertising magazines, credit facilities and mass consumption. Although the
critique of consumerism correctly points out that many aspirations in the
population cannot be adequately satisfied by consumer society, because, for
example, the unemployed do not possess purchasing power, it is also the case
that the content and nature of advertising are shaped and determined by
consumer needs. With changes in the nature of employment and the household, the focus of advertising has shifted from the young to the middle aged.
The relationship between needs and consumption is far more complex than
the hegemonic argument suggests.

The critique of capitalist consumerism has eventually to rest on some
notion of real needs and on some distinction between need and pleasure.
Desires are 'vain', but needs are 'real'; capitalism operates at the level of trivial pleasures, but it cannot, according to the consumer critique, ultimately
satisfy our needs. Behind this argument there is another assumption:
exchange-value is bad, use-value is good (Kellner, 1983). By virtue of our
embodiment, we have real and mundane needs which must be satisfied and
these needs are universal, which in some respects define what it is to be a
member of the human species. There are various problems with this position
which are explored throughout this study. There is, however, one point
which we can note immediately about the argument from universal needs.
What we can say about these needs is generally vague and trivial. Human
beings need to eat, but what, when and how they eat is entirely variable.
Individual variations in sleep patterns, sexual activity and eating habits
appear to be unlimited. Even our individual anatomy is variable (Williams,
1963). The problem is that we live in a socially constructed reality and our
pleasures are acquired in a social context, but this is also true of 'need'. To
some extent the contrast between 'need' and 'desire' is grounded in a distinction between 'nature' and 'culture'. Our needs are seen to be real,
because they are natural and they are natural because our bodies are a feature of the natural landscape of our existence. By contrast desires are vain,
because they are cultivated. Our culture emerges from the cultivation of
our bodies and the more civilized we become, the more unnecessary our cultural baggage appears to be. While desire is mere luxury, needs are
necessities. This distinction is difficult to maintain, because what we perceive
as needs are in fact thoroughly penetrated and constituted by culture. The
distinction between need and desire is primarily a value-judgement. In the
mediaeval period, theologians condemned husbands who found pleasure in
the bodies of their wives; in the twentieth century, critics of consumerism
condemn the middle class who find pleasure in vain commodities. Both critiques make a value-judgement based on a distinction between necessities
and luxuries. What we regard as a need is very much bound up with expectations about what is normal, and what is normal is not simply a statistical

criterion because what is normal is essentially cultural. The oddity is that in everyday language we often use 'normal' interchangeably with 'natural' and thus what conforms to nature is what conforms to social expectation. However, since with technological and social changes, modern societies are less exposed to and dependent on 'nature', nature as a criterion for social arrangements becomes increasingly irrelevant. Social change rolls back the barrier of natural necessities.

The ontological status of 'nature' is of particular importance in the debate about gender relations. There is general agreement in sociology that notions like 'maternal instinct' and 'maternal deprivation' are aspects of an ideology which induces women to stay at home as mothers. The conventional view of women as mothers confuses 'mothering' with 'parenting'. More generally, while there are biological differences between men and women, these are culturally mediated and historical. What we regard as male and female characteristics are socially constructed differences and these characteristics can be radically changed by social and political intervention. The logic of this argument would, however, also include the notion that biology is itself socially mediated and that biology is a classificatory system by which experiences are organized. What stands behind 'gender' is not an unmediated reality but another level of social constructs and classifications; the anatomy of the body is precisely such a classification (Armstrong, 1983). 'Gender' is a social construct which mediates another social construct of 'biology'. There are no natural criteria for judging what is valuable or real and to admit that there are biological differences between men and women may be perfectly admissible, but it necessarily means the adoption of a perspective. Biology is cognitive systematization (Rescher, 1979). Biological facts exist but they exist by virtue of classificatory practices which preclude fixed points (such as 'nature') precisely because we inhabit a world that is perspectival.

Concepts like 'desire', 'need' and 'appetite' are part of a discourse by which we describe rather than explain. From a structuralist perspective, 'biologism' is one type of discourse; 'feminism' is another. Structuralism regards these discourses as autonomous since 'texts' have a life of their own. Although structuralism represents a particularly powerful position, the argument of this study is that the discourse of desire, and more generally the location of the body, has to be understood in terms of massive changes in the whole structure of societies. The debate about the nature of women in modern societies is an effect of the changes in the social position of women and the transformation of the social role of women is an effect of the reorganization of capitalism. Whereas the economic process of feudalism required the detailed control of female sexuality within the landowning class, the organization of property in late capitalism does not require a regimen of sexual control. Capitalism no longer depends on the existence of the nuclear family and the structure of the household has changed fundamentally in the post-war period. The traditional notion that women were desirable but not desiring has collapsed along with the Victorian family and the double-standard. It is not inconceivable that capitalism will cease to be a society in which there are definite sexes; genetic engineering certainly makes this out-come technically possible. What contemporary capitalism does require is the security of production, a technology of consumption and the commercial legitimation of desire. The differentiation of bodies by sex is increasingly irrelevant to these three conditions.

# 2

# Sociology and the Body

## Absent Bodies

Contemporary sociology has little to say about the most obvious fact of human existence, namely that human beings have, and to some extent are, bodies. There exists a theoretical prudery with respect to human corporality which constitutes an analytical gap at the core of sociological enquiry. The collective phenomena of births, ageing and mortality have become the academic monopoly of historical and mathematical demography, where the moral and social significance of these events is subdued in favour of exact calculation. What one might term the theodicy of the body is equally neglected even in the sociology of religion (Turner, 1983). The oddity of the failure of sociology to develop a theory of the body and bodies is emphasized by the prevalence of commonsense notions that diet, jogging, fasting, slimming and exercise are not merely essential aids to sexual fulfilment, but necessary features of self-development in a society grounded in personalized consumption. Some recent debates in sociology, particularly with respect to narcissism (Lasch, 1979), have illustrated an awareness of the changing symbolic significance of the body in relation to capitalist development, but these are exceptions that prove the rule. The reasons for sociology's exclusion of the body from theoretical enquiry are not difficult to trace.

The epistemological foundations of modern sociology are rooted in a rejection of nineteenth-century positivism, especially biologism which held that human behaviour could be explained causally in terms of human biology (Parsons, 1937). Sociology emerged as a discipline which took the social meaning of human interaction as its principal object of enquiry, claiming that the meaning of social actions can never be reduced to biology or physiology. The academic institutionalization of sociology involved its separation from eugenics and Darwinist biology. It is clearly the case, however, that evolutionary biologism played an important part in the theoretical development of sociology, especially in the work of Herbert Spencer (Peel, 1971) and Patrick Geddes (Boardman, 1978). It can also be argued (Foucault, 1973) that the emergence of social science was closely connected with the growth of rationalized medicine, through the collection of health statistics with the growth of urban populations in the nineteenth century. Despite these institutional and theoretical connections with positivist biology and medical science, the central assumptions of sociology were inimical to its submersion in biology. The physical sciences and bastard offshoots like sociobiology do not provide a model for the explanation of social reality which cannot be subsumed in 'nature'. The central assumptions of sociology are that the natural world is socially constituted and transformed by human activity. Human beings do not simply apprehend the natural world as a given, since nature is always mediated by culture. In arguing that the reality in which the human species is situated is socially constructed (Berger and Luckmann, 1967), sociology has to some extent incorporated the argument of Karl Marx that man

> opposes himself to Nature as one of her own forces, setting in motion arms and legs, head and hands, the natural forces of his body, in order to appropriate Nature's productions in a form adapted to his own wants. By thus acting on the external world and changing it, he at the same time changes his own nature. (Marx, 1974, vol. 1, p. 173)

The external world, including the human body, is not a given, but an historical reality constantly mediated by human labour and interpreted through human culture. The human body as a limiting point of human experience and consciousness seemed less important than the collective reality of the social world within which the self was located. The legitimate rejection of biological determinism in favour of sociological determinism entailed, however, the exclusion of the body from the sociological imagination. The primary dichotomy of sociological theory was not Nature/Society, but Self/Society.

## The Self

Sociology extracted itself from the physical sciences as a model of social theory by viewing itself, according to Max Weber (1978), as an 'interpretative science' of the meaning of social action and interaction. Such interaction occurred between entities which were designated as 'the self' or 'the social actor' or 'the social agent'. The interaction of bodies is 'behaviour', whereas the interaction between social actors involves meaning and choice; it is the proper object of sociology. The social thus came to be seen as an on-going process of interactions between Ego and Alter, so that 'society' is an emergent reality and the product of ceaseless interactions. It is important to note that social actors (Ego and Alter) are not necessarily 'real' individuals, but socially constituted entities. For example, Alfred Schutz (1962) made an elementary distinction between direct, face-to-face interaction with consociates and indirect action with predecessors, successors and contemporaries. In sociology, it is perfectly reasonable to include in 'interaction' exchanges between the living and their dead ancestors, between children and their dolls, between the faithful and their gods. A 'social actor' is an entity which is socially constituted as an interactant. In the perspective of symbolic interactionism (Rose, 1962), interaction fundamentally presupposes, as it were, an internal conversion in which I interact with myself. The 'I' is the response of an individual in a total fashion to the variety of attitudes of others; the 'me' is the organized attitudes of others. The self is thus the complex union through interaction, symbol and gesture of the I and the me (Strauss, 1964). By concentrating on the self as a symbolically constituted phenomenon, symbolic interactionism reinforced the more widespread sociological perspective in which the corporality of social actors was relatively insignificant in social action. The self is fundamentally sociological not biological, since the self is little more than a principle for the organization of gestures. The idea that the body might be one component of the continuity of the self was discarded in favour of the argument that the continuity of the self rests on the continuity of others' perceptions of personal continuity. In this respect, symbolic interactionism aligned itself with a particular philosophical position on the traditional mind/body problem in which the persistence of identity does not depend on the continuity of the body but on the coherence of memory and consciousness. In summary, the emphasis in sociology on the socially constituted nature of social being resulted in an implicit position that the body of the social actor is a largely inconsequential feature of the self-in-society perspective.

It has been argued that one reason for the submergence of the body in sociological theory was an unintended consequence of a legitimate critique of biologism. In response, sociology emphasized the importance of culture and symbolism in the organization of the self and society. In the formation of sociology, however, the rejection of biologism is often difficult to dissociate from the rejection of methodological individualism and, more generally, from so-called 'atomism'. Although some sociologists have been fervent methodological individualists, whether covertly or overtly, the central tradition of sociology denies the argument that

> The ultimate constituents of the social world are individual people who act more or less appropriately in the light of their dispositions and understanding of their situation. Every complex social situation, institution or event is the result of a particular configuration of individuals, their dispositions, situations, beliefs, and physical resources and environment. (Watkins, 1959, p. 505)

The macro-sociological tradition has taken the social structure and the structure of collectivities as the constituents of society, arguing that 'structure' cannot be reduced to the relationship between individuals and that 'society' is *sui generis*. Because macro-sociology has, for example, been concerned with the relationship between social classes and political parties, between the state and the economic basis of society, and between the family and economic change, the human body cannot be located within this theoretical space. Whereas micro-sociology excludes the body because the self as social actor is socially constituted in action, macro-sociology excludes the body because its theoretical focus is on the 'social system'. In the latter tradition, any theoretical concentration on the body must smack of methodological individualism, since it is assumed on a commonsense basis that the individual is uniquely located in a body. Thus any attempt to direct sociology towards a theory of the body must appear as an heretical betrayal, since such a movement suggests simultaneously biologism and methodological individualism.

## Michel Foucault

In writing about sociology's neglect of the body, it may be more exact to refer to this negligence as submergence rather than absence, since the body in sociological theory has had a furtive, secret history rather than no history at all. The point of this book is to expose this submergence and to articulate a theory in order to bring out the prominence of the body and bodies. Before drawing attention to various areas of sociological theory where, so to speak, the body survived despite its theoretical exclusion, it is important to outline in a peremptory manner what is embraced by the sociology of the body. This brief recovery of the body has to be inserted in order to avoid any hasty accusations of biologism or atomism. Since this book, as will become evident later, is in part an application of the philosophy of Michel Foucault, some of the basic distinctions here are clearly Foucauldian. First, a sociology of the body can be regarded as a materialist enquiry. In an interview concerning power and the body, Foucault compared his own interest in the body with Marxist analysis of ideology and power:

> As regards Marxism, I'm not one of those who try to elicit the effects of power at the level of ideology. Indeed I wonder whether, before one poses the question of ide- ology, it wouldn't be more materialist to study first the question of the body and the effects of power on it. Because what troubles me with these analyses which priori- tise ideology is that there is always presupposed a human subject on the lines of the model provided by classical philosophy, endowed with a consciousness which power is then thought to seize on. (Foucault, 1980a, p. 58)

For Foucault, the power-effects of ideology are not to be seen in terms of the manipulation of the human subject as pure consciousness. In modern soci- eties, power has a specific focus, namely the body which is the product of political/power relationships. The body as an object of power is produced in order to be controlled, identified and reproduced. Power over the materiality of the body can be divided into two separate but related issues – 'the disci- plines of the body and the regulations of the population' (Foucault, 1981, p. 139). The first relates to singular bodies and is referred to as an 'anatomo-pol- itics', while the second embraces the species body and involves a 'bio-politics' of populations. Foucault regards medical science as the crucial kink at the level of knowledge between the discipline of individual bodies by profes- sional groups (of psychiatrists, dietitians, social workers and others) and the regulation of populations by panopticism (in the form of asylums, factories, schools and hospitals). The administered society involves the control of per- sons through the medicalization of bodies. While often presented as a critique of modern Marxism, Foucault's project can be seen to bear a relationship to a view of historical materialism presented by Friedrich Engels who, in *The Origin of the Family, Private Property and the State* (n.d), claimed that the materialist interpretation of history regarded the production and reproduc- tion of immediate life as the determining factor of human societies. This determination had a two-fold character, namely the production of the means of subsistence and the production of human beings: 'The social institutions under which men of a definite historical epoch and of a definite country live are conditioned by both kinds of production: by the stage of development of labour, on the one hand, and of the family, on the other' (Engels, n.d., p. 6). A materialist theory of the body has to provide the linkage between the discipline of the body and the regulation of populations in terms of the institutional connections between family, property and patriarchy.

While human society has changed fundamentally over the last 2000 years, sociobiology would suggest that the human body has remained, in all important respects, physiologically static. The implications of this juxtaposition are that a sociology of the body would be an historical enterprise. Such a conclusion is, however, fundamentally misguided, since the questions of 'the body' and 'the population' in relation to socio-cultural structures are necessarily historical. This insight into the historicity of the body is one of the basic contributions of Foucault's approach to the history of Man as an object to science. With the demographic explosion of the eighteenth century, 'population' emerged as an object of innumerable scientific technologies and enquiries:

> Within this set of problems, the 'body' – the body of individuals and the body of populations – appears as the bearer of new variables, not merely as between the scarce and the numerous, the submissive and the restive, rich and poor, healthy and sick, strong and weak, but also as between the more or less utilisable, more or less amenable to profitable investment, those with greater or lesser prospects of survival, death and illness, and with more or less capacity for being usefully trained. The biological traits of a population became relevant factors for economic management. (Foucault, 1980a, p. 172)

'Population' emerged as the focus of the sciences of the body and in association with new disciplines, regulations and coercive practices. With the conjunction of the body and the population, the sexuality of individuals became a new focus of power relations which were directed at a management of life.

## Spirit and Flesh

In more conventional terms, the body can thus become a genuine object of a sociology of knowledge. The Western tradition of the body has been conventionally shaped by Hellenized Christianity, for which the body was the seat of unreason, passion and desire. The contrast in philosophy between mind and body is in Christianity the opposition between spirit and flesh. The flesh was the symbol of moral corruption which threatened the order of the world; the flesh had to be subdued by disciplines, especially by the regimen of diet and abstinence (Turner, 1982a; 1982b). The body in Greek thought had been the focus of the struggle between form and desire (between Apollo and Dionysus). Christianity inherited this viewpoint, but darkened it by seeing the flesh as the symbol of Fallen Man and irrational denial of God. In mediaeval times, the celebration of the body in festival and carnival came to be a political expression of popular dissent against the dominant literate tradition of the court and the urban centres of social control. Rabelais's confirmation of the primitive and popular language of the body in the tradition of the market place and the carnival was thus an affront to the refinement expressed in 'official' literature (Bakhtin, 1968). Within the sociology of knowledge, therefore, it is possible to trace a secularization of the body in which the body ceases to be the object of a sacred discourse of flesh to an object within a medical discourse where the body is a machine to be controlled by appropriate scientific regimens. The history of this transition is complex. In gymnastic systems, the rationalization of movement represented an application of Borelli's iatrophysical school of medicine (Broekhoff, 1972). In dietary practices, there was a shift away from an eighteenth-century concern for long life as a religious value to the nineteenth-century concern for the efficient quantification of the body (Turner, 1982a). The result of these changes was to reify and objectify the body as an object of exact calculation.

The idea that the body is the location of anti-social desire is thus not a physiological fact but a cultural construct which has significant political implications. The contradiction between passion and reason as the basis of Durkheim's *homo duplex* is also the vindication of authority which provides the root of social order and social solidarity (Sennett, 1980). One of the principal arguments of this study is not simply that the body is culturally constructed in opposition to social authority, but specifically that the feminine body is the main challenge to continuity of property and power. The division between female passion and male reason is thus the cultural source of patriarchy. While patriarchy exists independently of the capitalist mode of production, being a specific distribution of power, capitalist society has articulated this division by providing a spatial distribution of reason and desire between the public and private realm, institutionalized by the divorce between the family and the economy.

In the ancient world, the private space of the domestic economy was the arena of necessity and deprivation, whereas the public sphere of the citizen was equated with freedom. Thus, the private space of the hearth was connected with the production of life's necessities by beings (slaves and women) who were not entirely human (Arendt, 1959). The growth of privacy as a value presupposes the development of a doctrine of the private individual, an ideology of familialism, an institutional separation of family and economy in which the domestic unit ceases to have productive functions, and a large bureaucratic apparatus by which the public life of individuals is measured and calculated for the purpose of social control. The union of capitalist industrialization, utilitarian individualism and the nation-state provided the general conditions for the rise of the division between the public and the private world. The important feature of this division in modern society is that the private space is characterized by the intimacy and emotionalism of the household which exists specifically for the servicing of the body, namely the production of children, socialization and the servicing of the labour force.

There is thus a sharp contrast between the formality, impersonalism, neutrality and universalism of work in public space, and the informality, particularism and affectivity of the private home. In the social division of society, there is also a sexual division by which certain activities ('mothering' and 'working') become gender specific. In addition we can suggest that there is a spatial division between passions (private sphere) and reasons (public sphere).

Table 1

| Private | Public |
| --- | --- |
| gemeinschaft | gesellschaft |
| desire | reason |
| female | male |
| informal | formal |
| affectivity | neutrality |
| particularity | universality |
| diffusion | specificity |
| hedonism | asceticism |
| consumption | production |

This can be illustrated as shown in Table 1. In making this distinction between private/public, it would probably be more accurate to refer to private spaces in the plural. The modern home is opened to the world by an architectural emphasis on light and space. At the same time the home remains a castle cut off from other private spaces. The transition from the Renaissance to the modern world thus involves a transition from the 'open' body linked to the public world through ritual and carnival to the 'closed' body of individualized consumer society (Bakhtin, 1968). Desires are now inscribed in private bodies separated from the hygienic space of the public world.

## Sociology of the Body

To write a sociology of the body is thus not to write a treatise on society and physiology. It involves the historical analysis of the spatial organization of bodies and desire in relation to society and reason. The principal contours of such a study may be stated as the following:

(1)   For the individual and the group, the body is simultaneously an environment (part of nature) and a medium of the self (part of culture). The body is crucially at the conjuncture of human labour on nature through the medium of writing, language and religion and thus critically at the conjuncture of the human species between the natural order of the world and the cultural ordering of the world. The transaction between nature and society can thus be seen in terms of the body as physiology (that is, an internal environment). To take one obvious example, the body has physiological needs, in particular food, liquid and sleep. The nature, content and timing of these activities of eating, drinking and sleeping are subject to symbolic interpretations and to massive social regulation. We can thus

think of the body as an outer surface of interpretations and representations and an internal environment of structures and determinations.

(2) Corresponding to the internal/external division, it is important to make a distinction between, following Michel Foucault, the body of populations and the body of individuals. It has been argued that in Western culture the site of desire is the internal body which has to be controlled by the rationalized practices of asceticism (such as religious fasting and the medical regimen). Similarly, the body of the individual is regulated and organized in the interests of population. The control of group sexuality is the most obvious illustration. No society leaves social reproduction to the free choice of individuals. While in modern industrial society sexual behaviour often appears as the free choice of the private consuming citizen, there are regulations relating to abortion, infanticide, illegitimacy, homosexuality and prostitution. The regulation of the body of populations occurs along two dimensions of time and space, that is the regulation of reproduction between generations and the regulation of populations in political/urban space. The sociology of the body is thus a political sociology, since it concerns the authoritative struggle over desire.

(3) The body lies at the centre of political struggles. While it can be argued unambiguously that the physiology of men and women represents a major difference (in reproductive functions), gender identity and gender personality have to be inserted into physiology by socialization into specific roles and identities. Similarly, while the body undergoes a natural maturation with ageing, the concepts of 'youth', 'infant', 'child' or 'senior citizen' are cultural products of historical changes in the organization of Western society (Ariès, 1962). The body – its character, structure and development – thus provides a basic metaphor of pre-modern social theorizing in such notions as the 'body politic', gerontology, gerontocracy, patrimonialism and patriarchy. For example, the debate about patriarchy, in its specifically political form, goes back to Sir Robert Filmer's *Patriarcha* (published posthumously in 1680). In this doctrine of patriarchal sovereignty, royal power is derived from divine power via Adam. Patriarchy rested on analogy. The king is the father over his kingdom; Adam was father over nature and humanity; God is father over man. Patriarchy thus comes before the authority of law and is the source of all rights and obligations. Authority is thus transubstantiated through the body of kings just as it is transubstantiated through the body of fathers. In religious systems, the authority of Christ is transubstantiated through the eucharistic elements of flesh and blood, just as in the political system the continuity of blood is essential to the continuity of power.

(4) Most forms of sociological theorizing make a sharp separation between the self and the body. G. H. Mead who in many respects provided the original philosophical basis of symbolic interactionism, wrote in *Mind, Self and Society* that

> We can distinguish very definitely between the self and the body. The body may be there and operate in a very intelligent fashion without there being a self involved in the experience. The self has the characteristic that it is object to itself, and that characteristic distinguishes it from other objects and from the body. (Mead, 1962, vol. 1, p. 136)

While the self/society contrast became the main focus of interactionist theory, it is also the case that most proponents of interactionism argue that the self is realized through performance. Crucial to self-performance is the presentation of the body in everyday life. It is possible therefore to reinterpret Goffman's sociology as not the study of the representation of the self in social gatherings but the performance of the self through the medium of the socially interpreted body. One important focus of his work is the question of the breakdown of the micro-social context through events which discomfort the self and social interaction. These include embarrassment and stigma. Significantly social disruptions are repaired by 'face-work' which reassert the normality of interaction. These disruptions of interaction are typically, but not exclusively, focused on the body – flushes, tears and stigmatic abnormalities. The body is thus crucial to both the micro and macro orders of society. The body is the vehicle for self-performances and the target through rituals of degradation of social exclusion. Intimacies and exclusions focus on the body as the means of indicating the self (Garfinkel, 1956; Weitman, 1970). A sociology of the body would thus also have to embrace a sociology of deviance and control, since mortifications of the self are inextricably bound up with the mortifications of the body. Again it would be appropriate to distinguish between the deviance of body surfaces (blushes, flushes, unwanted excreta) which are subject to cultural surveillance and those 'deviances' of the inner body (disease and illness) which are likewise objects of moral evaluation. The sociology of the body as vehicle of information about the self would thus divide around the stigmatology of the outer surface and a teratology of deformed structures.

A sociology of the body is not sociobiology or sociophysiology. It is not reductionism, although it is genuinely and literally a materialist analysis. As I shall elaborate later, a sociology of the body is a study of the problem of social order and it can be organized around four issues. These are the reproduction and regulation of populations in time and space, and the restraint and representation of the body as a vehicle of the self. These four issues presuppose the existence in Western society of an opposition between the desires and reason, which I have suggested articulates with a further set of dichotomies, especially the private/public, female/male dichotomies. The control over the body is thus an 'elementary' 'primitive' political struggle. The sociology of the body is consequently an analysis of how certain cultural polarities are politically enforced though the institutions of sex, family and patriarchy. This institutionalization is itself subject to certain major transformations of society (for example, from feudalism to capitalism) and the saliency of the four dimensions (reproduction, regulation, restraint and representation) is historically conditioned.

**Locations for a Theory**

While sociology has not overtly incorporated a sociology of the body, it has inherited the classical Western dichotomy between desire and reason which has informed much recent debate within sociological theory. This implicit theory has not been adequately or systematically examined. Crudely speaking, we can divide social philosophy between one tradition which treats nature/body/desire as the source of value and happiness in opposition to society/technology/reason, and a second tradition which regards desire/pleasure/the body as the negation of human value located in the life of the mind. My argument is that, mainly implicitly, sociological theory has been shaped by the opposition: civilization versus desire. As Daniel Bell has noted:

> The rational and the passionate – these are the axes around which social thinkers have organized their conceptions of human nature since the dawn of philosophy. But which is to prevail if men are to be just and free? For the classical theorists, the answer was plain. (Bell, 1980, p. 98)

That answer was the necessity of subordinating passion to reason in the interests of social stability and social order: Apollo over Dionysus. While that polarity has characterized Western philosophy from Plato onwards, the debate about passions received a significant impetus in the nineteenth century following the opening of a new discourse on sex in the late eighteenth century. First there was the Marquis de Sade (1740–1814), whose work has recently attracted considerable reappraisal (Barthes, 1977; Carter, 1979; de Beauvoir, 1962); the second was the neglected Charles Fourier (1772–1837). For Fourier, civilization stood in opposition to passion and, by imposing artificial social duties on desire, destroyed the 'natural' liberty of the subject:

> All these philosophical whims, which are called duties, have no relationship with nature. Duty comes from men; attraction comes from God; and to understand the designs of God it is necessary to study attraction, nature by itself without any reference to duty. . . . Passionate attraction is the drive given us by nature prior to any reflection; it is persistent despite the position of reason, duty, prejudice, etc. (Beecher and Bienvenu, 1972, p. 216)

Fourier is often claimed as a formative thinker within the socialist tradition (Kolakowski, 1978); Marx was, for example, sympathetic towards Fourier's economic analysis, but in general Fourier's emphasis on sexual liberation was not incorporated within Marxist thought. Neo-Marxism and critical theory have been forced to incorporate a modified Freudianism in order to be able to analyse the relationship between sexuality and society. This turn to Freud was especially evident in the work of Herbert Marcuse (1969). The materialist tradition of the nineteenth century largely rejected Fourier's utopia, but maintained his sharp dichotomy of desire and reason. Furthermore, Marx's concept of active materialism in the notion of labour did not attempt to deal with materialism as physiology.

The major attempt to resolve the mind/body dichotomy in the nineteenth century which provided part of the background to Marxism came from Ludwig Feuerbach (1804–1872). In his later work, Feuerbach attempted to solve the traditional puzzle of mind and body through the idea of sensibility. Feuerbach attempted to give this idea of sensibility a materialist basis by grafting it onto the digestive theories of Moleschott's *Theory of Nutrition* of 1850. The unity of thinking and being in the exchange between man and nature was located in man's appropriation of nature by eating. While Feuerbach assumed that the traditional riddle of materialism and idealism had been solved by the chemistry of digestion, which he summarized in the slogan 'Man is what he eats', Feuerbachian man, as Marx and Engels recognized, remains passive. Feuerbach failed to extend his 'diet-materialism' by recognizing that 'The dialogue between my stomach and the world in real activity, is mediated by the dialogue between production and consumption, the social dialogue of human praxis that Marx developed in his political economy' (Wartovsky, 1977, p. 416). While Engels dismissed Jakob Moleschott as a 'vulgar materialist' in the *Dialectics of Nature* (Engels, 1934), he regarded Feuerbach as an idealist, partly because Engels saw Feuerbach's intention as, not to replace religion, but to perfect it through anthropology. Feuerbach's philosophy remained idealist because it had no genuinely historical dimension and his thought was limited despite his attempt to resolve the classical problems of philosophy in the developments taking place in chemistry:

> For we live not only in nature but also in human society, and it too no less than nature has its historical development and its science. It was therefore a question of bringing the science of society, that is, the totality of the so-called historical and philosophical sciences, into harmony with the materialist base, and of reconstructing it on this base, but this was not granted to Feuerbach. (Engels, 1976, p. 25)

There is an irony in this rejection of physiology as the basis of materialism. Given this hostility to physiology, the question of the human body and its relationship to production and reproduction via the institutions of the family and patriarchy largely disappeared from Marxist philosophy. The main exception to this assertion is to be found in the work of the Italian Marxist, Sebastiano Timpanaro who in *Sul Materialismo* (1970) argued pessimistically that in death nature has its final and irreversible triumph over man. The problem of the body was submerged by this rejection of physiological materialism, which was regarded as ahistorical and passive. At the same time, Marx rejected the argument of Malthus and Malthusians that population pressures had a major importance for the analysis of economic growth and prosperity. The population issue had to be rendered as an historical question, not as a static restraint on the economic base: 'Every special historic mode of production has its own special laws of population, historically valid within its limits alone' (Marx, 1974, vol. 1 p. 693). For Marx, the idea that the accumulation of capital could be explained by reference to the control of sexual urges was a myth of hypocritical bourgeois theorists. Despite Marx's perceptive criticisms of the static nature of physiology as a basis for materialism, the consequence of these rejections was that Marxism did not, despite appeals to the notion of 'dialectics', address itself to the classical desire/reason problem. Furthermore, as a science, Marxism tended to embrace technical rationality. Consequently any interest in emotions, passions and desire, on the one hand, or populations and reproduction, on the other, was either diminished or seen to be the result of heresy, especially methodological individualism. Contemporary theoretical interest in the body/desire couple has thus been primarily stimulated by debates with Freudianism, which have emerged in two wings of modern social theory – critical theory and structuralism.

## Critical Theory

The early work of the Frankfurt School (Jay, 1973) saw the struggle by man to dominate nature through technical rationality as resulting in political slavery and the renunciation of feeling. This theme can be seen explicitly in Adorno, especially in Horkheimer and Adorno, *Dialectic of Enlightenment* (1973) where they explored the myth of Odysseus and the Sirens. Odysseus avoided the temptation of the Sirens' songs by blocking the ears of his sailors with wax and by lashing himself to the mast. This myth represents the psychological logic of bourgeois civilization in which the workers have to deny and sublimate their emotions in favour of hard work and practicality, while the bourgeois capitalist must restrain and discipline desire in the interests of further accumulation. Enjoyment through consumption stands in the way of economic growth; capitalism requires the control of nature through technology but also requires, as it were, the control of inner nature in the human species. Because personal ecstasy is 'a promise of happiness which threatened civilization at every moment' (Horkheimer and Adorno, 1973, p. 33), civilization was seen by critical theorists in terms of renunciation. It is thus easy to understand why the Frankfurt School perceived the relevance of Freud's analysis of the permanent conflict between egoistic pleasures and social controls. However, for writers like Marcuse, capitalism at least had the economic potential to satisfy basic needs and to reduce the social requirement for psychic repression. The realization of that potential became the main political battle within late capitalism, a battle which aims 'to minimize the self-destructive aspects of human desires' (Leiss, 1972, p. 197).

Critical theorists, especially Marcuse, came to see hedonism as a potentially liberating force in society. Classical hedonism protested against the view that happiness was essentially spiritual by demanding 'that man's sensual and sensuous potentialities and needs, too, should find satisfaction' (Marcuse, 1968, p. 162). The failure of the Cyrenaic version of hedonism was that it took wants and needs as empirically given, and its commitment to ethical relativism prevented hedonism from making judgements about true and false happiness, or between short-term and long-term pleasures. This version of hedonism ruled out any critique of capitalist society the existence of which depends partly on fostering false needs through advertising and mass consumption. By contrast Epicurean hedonism attempts to differentiate between pleasures with the aid of reason. Marcuse thus suggested that the traditional opposition between reason and desire was false, since, in the case of Epicurus 'reason is made a pleasure' and 'pleasure is made reasonable' (Marcuse, 1968, p. 171). Capitalism, however, involves the splitting of reason and pleasure by restricting pleasures to the sphere of consumption and harnessing reason to the needs of technical production.

For Marcuse, classical Marxism is increasingly redundant as a theory of late capitalism. Marx could not and did not fully grasp the liberating potential in automation which could in principle free human labour from conditions of drudgery and boredom. In addition, Marcuse suspected that Marx's emphasis on labour harboured a puritanical, moralistic attitude towards play and leisure as mere epiphenomena. Against Marxism, Marcuse suggested that sexual fulfilment would result in a liberating devaluation of work and labour. In many respects, this is a distinctively odd view of Marx, since Marx in the manuscripts of 1844 in many respects precisely anticipated the view of critical theory that civilization equals renunciation. Indeed, these were his precise words:

> Political economy, this science of wealth, is therefore simultaneously the science of renunciation, of want, of saving – and it actually reaches the point where it spares man the need to either fresh air or physical exercise. This science of marvellous industry is simultaneously the science of asceticism, and its true ideal is the ascetic but extortionate master and the ascetic but productive slave. Its moral ideal is the worker who takes part of his wages to the savings-bank, and it has even found readymade an abject art in which to embody this pet idea . . . political economy – despite its worldly and wanton appearance – is a true moral science, the most moral of the sciences. Self-renunciation, the renunciation of life and of all human needs, is its principal thesis. (Marx, 1970, p. 150)

The difference, however, between Marx and Marcuse is that in Marx the idea of human happiness as a criterion of social progress is relatively unimportant by comparison with other values, like freedom and equality. One other difference is that Marcuse reifies and unifies the emotional life of people into ahistorical oppositions between Reason and Nature or Man and Desire. While Marx and Engels in *The German Ideology* (1974) criticized the 'speculative philosophers', such as Feuerbach, for converting the life of real, sensuous men into the abstract 'essence of man', Marcuse persistently abstracts people from their social relations in order to write about the hedonistic interests of Man. Both aspects of Marcuse – the elevation of happiness to a political value and the reification of Man – have been usefully criticized by MacIntyre 'in making "Man" rather than "men" the subject of history he is at odds with Marx and that in making "happiness" a central goal of man's striving he is at odds not only with Hegel, as Marcuse himself recognizes, but also once more with Marx' (1970, p. 41). Two criticisms of Marcuse's version of critical theory follow from this observation by MacIntyre. First, if we are to avoid any notion of a unified 'essence of man', then it is important to avoid even the generic notion of 'men'. If the emotional life of people is fundamentally bound up with the particular social relations in which they are embedded, then we can have no unified concept of pleasure. We can only talk of 'pleasures' which are specific to particular persons in particular contexts. The implication is that 'pleasures' are inevitably relativistic, idiosyncratic, peculiar and personal. If this is the case, then Marcuse's search for some notion of universally valid hedonism which is compatible with reason is false. In other words, the Epicurean version of hedonism fails, because no universal standard of critical reason could adjudicate *my* pleasures. In the last analysis, I am the only authority on my pleasures. My preference for anti-social, short-term pornography may be incompatible with critical theory and the product of capitalist exploitation, but this preference is still pleasurable. One reason for this personal authority is that the relationship between my pleasure and my body is irreducibly immediate and intimate. This observation leads to the second criticism of Marcuse, namely that, despite all the talk about sexuality, Marcusean pleasures are strangely dissociated from the body. While it is obvious that thinking and imagining are activities we would describe as pleasurable, most of our pleasures involve the body because these pleasures typically involve physical sensations – eating, sleeping, sexuality, exercise, resting. I am not arguing that these are simply physical activities. They are in fact deeply cultural or at least mediated by culture, but they also presuppose that people have bodies and that the person is embodied. Marcuse does not take seriously Marx's observation in the theses on Feuerbach that sensuousness is 'practical, human-sensuous activity'.

**Structuralism**

In modern social theory, it is pre-eminently in the work of Michel Foucault that the human body is located centrally as an issue of knowledge. The importance of the body and desire in modern structuralist thought has often been recognized (Benoist, 1978), but the question of the body has a peculiar persistence in Foucault's approach to historical analysis. His ideas are difficult to grasp, but one important feature of his perspective is that, while in most conventional philosophy and social theory power is seen to repress desire, Foucault treats power as constructive and productive: desire is brought about by power/knowledge. While modern societies often appear to be characterized by sexual repression, in fact sexuality is constantly produced and examined by contemporary discourses, but these have come under the control of medical and psychiatric professions. The will to know has become the will to know sexuality and, since to know is to control, the sexual body has become the specific object of politics (Lemert and Gillan, 1982). There is, therefore, a very real difference between the approaches of Marcuse and Foucault in relation to the repression and representation of sexuality. For Marcuse, the repression of sex in capitalism is real and constitutes part of the surplus repression of libidinal pleasures. For Foucault, sexual repression is a myth, since sex has in fact become the object and product of endless scientific discourses – psychoanalysis, demography, biology, medical science – which aim to control and normalize sexuality. Knowledge produced desire in order to control it. In this respect, Foucault avoids the pitfall of treating desire as a unified phenomenon in history precisely because he treats desire as the product of certain historical discourses. However, this creates an ambiguity in Foucault's theory. At times he treats the body as a real entity – as, for example, in the effects of population growth on scientific thought or in his analysis of the effect of penology on the body. Foucault appears to treat the body as a unified, concrete aspect of human history which is continuous across epochs. Such a position is, however, clearly at odds with his views on the discontinuities of history and with his argument that the body is constructed by discourse. This, one interpretation of Foucault asserts that

> Clearly Foucault does not adopt Merleau-Ponty's solution. The body of desire is not, for him, the phenomenal, lived body. It is not a corporeal, incarnate subjectivity. . . . Desire, for Foucault, is neither expressed in the body, nor is the body the lived form of desire. (Lemert and Gillan, 1982)

On the other hand, Foucault has also said that, rather than starting with the analysis of ideology, it would be 'more materialist to study first the question of the body and the effects of power on it' (Foucault, 1981, p. 139). Such a materialist project would appear to take the corporeality of life seriously. What is 'the body'? is thus a question which is central to Foucault's thought, but one which is not clearly answered.

Foucauldian structuralism is, at one level, a response to Cartesian rationalism. By splitting people into body and mind, Descartes represents an important stage in Western thought. The Cartesian revolution gave a privileged status to mind as the definition of the person ('I think, therefore I am') and an underprivileged status to the body which was simply a machine. To some extent, Foucault reversed this situation by denying any centrality to subjectivity (the thinking, Cartesian subject) and by treating the body as the focus of modern discourse. Having rejected the transcendental Subject as merely a modern substitute for God or Logos, Foucault appears reluctant to have the Body as a controlling centre of social theory. The body is thus problematic for his theory. It looks as if Foucault wants to write the history of discourses about the body, of how the body is theoretically constructed, but this is specifically denied when he claims not to be producing a 'history of mentalities' which 'would take account of bodies only through the manner in which they have been perceived and given meaning and value; but a "history of bodies" and the manner in which what is most material and most vital in them has been invested' (Foucault, 1981, p. 152). To some extent, part of these difficulties is a product of his prior commitment to certain epistemological problems and thus the difficulties may be somewhat artificial. To reject Cartesianism, it is not necessary to deny the corporeal nature of human existence and consciousness. To accept the corporeality of human life, it is not necessary to deny the fact that the nature of the human body is also an effect of cultural, historical activity. The body is both natural and cultural.

## Foucault and the Origins of Sociology

Foucault's approach to the history of ideas has major implications for the sociology of knowledge, but specifically for the history of sociology. Foucault has rejected the conventional view that sociology had its origins in French positivism:

> Countless people have sought the origins of sociology in Montesquieu and Comte. That is a very ignorant enterprise. Sociological knowledge (*savoir*) is formed rather in practices like those of the doctors. For instance, at the start of the nineteenth century Guepin wrote a marvellous study of the city of Nantes. (Foucault, 1980a, p. 151)

The rise of modern medicine was associated with the development of new bureaucratic techniques in the panopticon system, the utilization of social surveys to map the distribution of diseases, the adoption of clinical methods for case-records and the elaboration of societal surveillance. Modern medicine is essentially social medicine as a policing of populations and a clinic of bodies. Sociology has it origins, along with social medicine, in the knowledge and control of populations which survey techniques made possible. The implication of Foucault's view of the birth of the clinic (1973) is that medical sociology as the study of the health of populations and of the body of individuals is central to the sociological enterprise as a whole and that sociology cannot be divorced from medicine. This view runs counter to the conventional interpretation of medical sociology which treats the sub-discipline as a late addition to the sociological curriculum. Most textbook introductions to medical sociology locate its institutional origins between 1955 and 1966 (Cockerham, 1982) and argue that medical sociology has not developed theoretically because of its subordination to the management interest of professional medicine (Roth, 1962; R. Strauss, 1957). The implication of Foucault's perspective is that sociology is applied medicine and its target is the regulation of bodies.

To some extent, this interpretation of the origins of sociology was anticipated by the notion of 'clinical sociology' which was first explicitly used by Louis Wirth (1931), who thought that sociologists would come to play a major role in the work of child-guidance clinics and who anticipated the spread of 'sociological clinics'. The value of sociology for medicine was its perspective on the 'whole person' whose illness could only be understood within a total social context. Wirth's views on clinical sociology were also echoed by L.J. Henderson who argued that sociology should adopt the clinical technique of medicine as a model for social observation and that medicine had failed to grasp the significance of the doctor–patient relationship as a social system (1935). It is interesting to note, given Foucault's comments on medicine and the origins of sociology, that Henderson saw 'the practice of medicine as applied sociology' (1936). It was Henderson's view of this proximity between medicine and sociology that provided the immediate context for Parsons's analysis of the 'the sick role' (1951), an analysis that formed much of the basis for modern medical sociology. Some recent studies from a Foucauldian perspective on the social role of the clinic, the dispensary (Armstrong, 1983) and medical orientations to 'the whole person' (Arney and Bergen, 1983) appear to duplicate this earlier emergence of clinical sociology.

Although medical sociology can be criticized for being merely an applied sociology whose aim was to facilitate the patient's compliance to the medical regimen, it was an area of sociological investigation which could not wholly avoid the problematic relationship between nature and culture. The debate about the 'sick role' kept alive the ambiguous nature of 'illness' and 'disease' as cultural categories (Mechanic and Volkart, 1961); it also provided a site in sociology where the critique of the medical model could be effectively located (Veatch, 1973). Because medical sociology is ultimately about the problem of social ontology in a very specific manner, it constantly raises questions about the status of the embodiment of human beings and it is therefore a theoretical location for a sociology of the body. The importance of Foucault's work on medical history is that it has made the theoretical nature of medical sociology more obvious; at the same time it has alerted us to the historical and political linkage between medicine and sociology. A sociology of the body is thus fundamentally an exercise within medical sociology.

## Phenomenology

It can be argued that structuralism had played a part in the modern analysis of the body either directly in the rejection of Cartesian dualism or indirectly in the analysis of the body as metaphor. Mikhail Bakhtin (1968) provided a rich diagnosis of the positional imagery of the body in mediaeval folk-humour; another illustration would be Roland Barthes (1973) in his analysis of the messages of striptease and wrestling. Structuralism was in part a rejection, therefore, of the presuppositions of rationalism, which were grounded in the Cartesian formula – cogito ergo sum. This rejection of the mind/body dichotomy was not peculiar to French structuralism, but was a position which characterized post-war French philosophy generally. For example, within the phenomenological movement (Spiegelberg, 1960) writers like Gabriel Marcel in his Le Mystère de l'Etre (1951) treated the body as the core of the ontological problem. Marcel argued that the body does not have a contingent or exterior relationship to existence, since my body is always immediately present in experience. He rejected the conventional dichotomies of subject/object and being/having to argue for the unity of mental and physical experience. For Marcel, to have a body is in fact always to be embodied so that existence is experienced-embodiment. The body is not an object or an instrument; rather I am my body, which is my primordial sense of possession and control. My body is the only object in which I exercise immediate and intimate rulership. For Marcel, therefore, the body is the ultimate starting point for any reflection on being and having, on existence and possession.

The mind/body legacy of Cartesian philosophy was also fundamental to the early philosophy of Jean-Paul Sartre, especially in *Being and Nothingness* (1957). To some extent, Sartre intensified the Cartesian division of mind and body by emphasizing the importance of intentionality of knowing (Danto, 1975). Under the influence of the phenomenology of Husserl and Heidegger, Sartre distinguished between being-in-itself (*en-soi*) and being-for-itself (*pour-soi*) in order to bring out the irreducible presence of free will and intentional action as necessary features of human existence. Because we experience freedom and responsibility as burdens, we are inclined to act as if our lives are determined by forces, whether psychological or sociological, which are beyond our control. We live, that is, in bad faith. The central doctrine of existentialism is that a person is essentially what they choose to be and to know (Warnock, 1965). Given the centrality of intentional consciousness in Sartre's existentialist philosophy, it might appear that the body has little part to play in our being-in-the-world. The problem of the body does, however, play an important part in Sartre's treatment of the philosophical question of the existence of other minds in his analysis of being-for-others. For Sartre, the body is our contact with the world which constitutes our contingency. Briefly, his argument is that we do not know other minds, but only minds as they are apprehended through the body. Sartre's account of the body is thus closely connected with his emphasis on intentionality and this feature of his argument is illustrated by his distinction between the three ontological dimensions. First, he drew attention to the body-for-itself. The body is not just a physical fact for me, alongside other facts – this typewriter, this chair or this paper – because my lived experience in the world is always from the point of view of my body. In seeing the world, I am not conscious of my eyes but only of a field of vision; my body-for-itself cannot be an object to me precisely because I am it. Furthermore, insofar as I apprehend my body at all, it is through objects in the world which indicate my location. My embodiment is indicated by the typewriter in front of me and the chair underneath. Secondly, Sartre distinguished the ontological dimension of the body-for-others. Whereas I cannot apprehend my body as an object but only as a body-for-itself, I apprehend the body of the other as an object about which I take a point of view and realize that my body as an object is the body-for-others. I do not perceive, however, the other's body as mere flesh, but always in a specific and concrete situation which I interpret as meaningful. The other is perceived not as a cadaver, but as a being-in-body with intentions whose actions or gestures are goal-directed and purposeful – such as striking a match to light a fire in order to eat. This interaction of my body as a subject for myself and an object seen by the other leads to the third ontological dimension. Being seen and observed by the other results in a recognition of my facticity, that I am an object to the other. In interaction, I begin to experience my intimate inside as an impersonal outside. The body-for-itself becomes objectified and alienated. What is my body is, through being observed by the other, simply a body.

Sartre's attempt to transcend Cartesian dualism has been criticized on a variety of grounds – for example by Merleau-Ponty in *Phenomenology of Perception* (1962) – but these objections may be summarized in the claim that Sartre did not overcome dualism: the problem was simply transferred to a distinction between *en-soi* and *pour-soi*, a distinction which is problematic and inconsistent with Sartre's commitment to an intentional ontology. The consequence of these criticisms and debates in French phenomenological philosophy is a rejection of any dualism between mind and body, and a consequent insistence on the argument that the body is never simply a physical object but always an embodiment of consciousness. Furthermore, we cannot discuss the body without having a central concern for intentions: the objective, 'outside' world is always connected to my body in terms of my body's actions or potential actions on them. To perceive the world is to reflect upon possible actions of my body on the world. Similarly, I experience my body as mine through my intimate, concrete control over my body. The basic idea of embodiment is that my animate organism

> is me, and *expresses* me: it is at once the self-embodiment of my psychic life, and the self-expressiveness of my psychic life. Thus, we can say, the problem of the experience of the body is the problem of embodiment. . . . The phenomenology of the animate organism is, accordingly, the descriptive–explicative analysis of the continuously on-going automic embodiment of consciousness by one organism singled out as peculiarly 'its' own, and, at higher levels, graspable by me as 'my own'. (Zaner, 1964, p. 261)

This view of the body from a phenomenological perspective is particularly important for sociology and, as I demonstrate in later chapters, especially for medical sociology as a critique of behaviourism. While the body is an object with specific physiological characteristics and thus subject to natural processes of ageing and decay, it is never just a physical object. As embodied consciousness, the body is drenched with symbolic significance. Phenomenology is a critique of behaviourism which, in treating the body as an object separate from consciousness, has to embrace, however covertly, Cartesian dualism. While the phenomenological critique is important, it is also limited as a philosophical basis for a sociology of the body.

The phenomenology of the body offered by Marcel, Sartre and Merleau-Ponty is an individualistic account of embodiment from the point of view of the subject; it is consequently an account largely devoid of historical and sociological content. From a sociological point of view, 'the body' is socially constructed and socially experienced. Their descriptive analysis of embodiment is of course consistent with phenomenological methodology which seeks to bracket out the question of existence in order to focus on the problem of meaning. Such an approach, however, brackets out too much, since in their own terms being involves meaning and vice versa. In presenting a sociological critique of their approach to embodiment, it is valuable to start with a consideration of the question: what is a person? The reason for this question is that the problem of the body in philosophical debate cannot be separated from the related issues of personhood, individuation and identity.

## The Person

In philosophical terms, an individual person in the full sense of the term is a being with a body, consciousness, continuity, commitment and responsibility. Some aspects of this ensemble of characteristics are contained in the following statement of what is implied in the notion of the identity of a human being:

> (1) the perception of an overall *coherence* – either 'substantive' or 'methodological' – within the experiences and expressions of an individual; (2) the memory of this individual and, normally, in at least some others of the *continuity* of the 'story' or 'tale' of his life; and (3) a conscious, but not wholly conscious, *commitment* to a particular manner of both comprehending and managing one's own self. (Kavolis, 1980, p. 41)

Each of these elements presents, admittedly, a number of difficulties. There has been much debate in philosophy as to whether individual persons are basically mental or physical entities (Shoemaker, 1963; Strawson, 1959). Whatever the philosophical problems, it is clear from a sociological stance that having a body with specific features, which has a particular placement in society, is crucial for everyday recognition and identification of persons. The interpellation of persons is typically the interpellation of specific bodies. This claim is not to deny that there are mistaken identities, false identities, impersonation and mimesis. The possession of a body is, despite these problems, an essential feature of the routine social identification of separate persons.

However, what both philosophy and sociology take for granted in the discussion of the body is that sociologically 'the body' is an animate organism. In fact the social notion of 'a body' is wider and more complex than this physicalist model implies. There are, for example, fictive and corporate bodies, at least some of which are also 'persons'. In mediaeval political theory, for example, it was held that kings had two bodies, one real and corruptible, one fictive and immortal (Kantorowicz, 1957). The king's sacred body was symbolic of the coherence and continuity of the whole society; the king's person embodied the body politic so that regicide was an attack on the person of the king and on the society as a whole. From a sociological point of view, it makes sense for a person to have two bodies, since the body is both a thing and a sign. In jurisprudence, the problem of identity has played a major part in the analysis of legal corporations as both bodies and persons. The idea of legal persons as legal unities, which were constituted as separate persons, was developed in the fourteenth century when Italian legal theorists were forced to deal with the emergence of corporations (Canning, 1980). While the law could conceive of human persons with rights and obligations, the law had not been fully developed to cope with collective entities like cities or trading corporations. To embrace such entities, the legal theorists developed the notion of *persona ficta* which was eventually developed into *persona universalis* – one person composed of many. In part, the theory of the corporation was seen to be analogous to the theory of the universal church. The real church was simply an embodiment of the universal church which in turn was simply the mystical body of Christ and as such continual and indestructible. While individual members of the physical church on earth were constantly replaced, the universal church had a continuity and existence independent of its actual adherents. The corporation was thus seen in a similar light; changes in individual membership did not influence the legal continuity of the whole. Hence, legal theorists conceived of the corporation as *persona perpetua*, such that membership of it changed the isolated nature of individual members (*homo separatus*) into corporate persons. In socio-historical terms, 'the body' is not necessarily the individual animate organism, because what will count as a body is an effect of social interpretation.

The idea that the individual person possesses self-consciousness is equally an historical notion. As Mauss (1979) has shown, the idea of the unity of body, person and consciousness is the result of a protracted historical process. The concept of 'person' comes from *persona*, a mask which was external to the individual. When persona in Roman law came to equal the 'self', it still excluded slaves who did not own their bodies, had no personality and had no claims over property. The notion of *persona* as a moral fact was first elaborated by the Stoics, but it was not until the development of Christian moral thought that the self was given an adequate metaphysical basis in the indestructibility of the unique soul. From Christianity, we derive the modern notion that the person is equivalent to the self and the latter is equivalent to consciousness. Much philosophical debate about what it is to be a person is thus culturally ethnocentric since there are cultures, such as classical Greece, in which human beings are not persons since they do not possess their bodies and have no public/legal identity.

There is also the argument that to have a personal identity as a separate individual is to have a certain coherence and continuity both from the point of view of the self and from others. What I am is bound up with my ability to recognize myself and the continuous recognition of myself by others. In everyday practice, being the 'same' person is measured by the way in which that person adheres to the 'same' beliefs, attitudes and practices over time in different situations. We are stigmatized for our inconsistencies and praised for our coherence. However, we also recognize that changes of identity and person are not wholly uncommon and indeed some cultures, like Protestant Christianity, regard the person as highly convertible. The 'true' person emerges only after religious conversion when the old Adam is destroyed. In other cultures, initiation rites (generally *rites de passage*) are held to produce new persons (Eliade, 1958) and to mark the change the body is often inscribed with symbols of such transformations (Brain, 1979). The convertibility of persons throws doubt on the view that continuity is essential to identity in the sense that our personal continuity may be in fact the product of continuity in social definition of the person/body unity.

The crucial issue in the debate about personhood is, however, that human persons are regarded as entities which are bearers of rights and responsibilities; to be a human person is to be capable of rational choice and consequently to be held responsible for our actions. We have seen that this argument was central to Sartre's existentialism which, given the importance of intentionality, regarded free will as an essential component of what it is to be a person. We differentiate between people and animals on a variety of criteria – language, rationality and a capacity for symbolization. One crucial feature is that we do not morally or legally hold animals responsible for their actions. A tiger which kills a man is not blamed for actions which are regarded as instinctual; a man who kills a tiger is blamed for actions which endanger a protected species. Furthermore, a person cannot be excused by saying 'my body did it' because we are thought to have intimate rulership, to follow Husserl, over our bodies. However, this question of responsibility and control over our bodies raises particularly difficult issues for phenomenologists.

The phenomenologists argue that persons have direct government over their bodies and that this regime is exercised without reflection: I do not have to tell my arm to move the carriage on the typewriter. Gabriel Marcel did, of course, admit that slaves have less control over their bodies than free citizens. However, the individualistic nature of phenomenology prevents it from developing a systematic theory about the social structure which unequally distributes the government of the body. The system of slavery is the most obvious example, but it can also be argued that under conditions of patriarchy women do not control their bodies because on entry into marriage they cease to be legal persons. The legal regulations known as *coverture* specify that the legal personality of the woman is merged with that of the husband so that the male head of household controls the bodies of his subordinates. The examples of intimate rulership of the body supplied by phenomenologists are trivial: lifting my pipe, picking up a glass or reading the newspaper. Less trivial is the fact that women under patriarchy do not control their sexuality because they do not make decisions about reproduction. Women, children, slaves and the insane do not in any important sense govern their bodies because they are denied full citizenship and are partly excluded from the public domain. It is over this issue that Foucault's analysis of sexuality can be seen to be powerful. A society which treats sexual freedom as a value in fact forces us to confess fully our inner 'secrets', especially to medical and paramedical experts: our freedom forces us to conform to standards of personal exposure. In addition we live in a world where our bodies are increasingly subject to inspection and surveillance by professional, occupational or governmental institutions. To talk about our phenomenological rulership of our bodies is to miss the crucial sociological point, namely the regulation of the body in the interests of public health, economy and political order.

**Summary**

The problem of the body lies at the juncture of the major issues of sociological theory. The epistemological problems of sociology, from the neo-Kantian movement onwards, centre on the dual membership of the human species in nature and culture. The human body is subject to processes of birth, decay and death which result from its placement in the natural world, but these processes are also 'meaningful' events located in a world of cultural beliefs, symbols and practices. At the individual level, my body is an environment that is experienced as a limit, but my consciousness also involves embodiment. I both have and am a body. Part of this distinction can be illustrated by the difference between disease and sickness. We might describe a person with Legg-Perthes' disease as suffering from degeneration of the thighbone, in which case the thighbone is diseased but not sick. Similarly we may describe an apple as diseased but not ill. Concepts like 'illness' and 'sickness' are socio-cultural categories which described the condition of persons rather than their flesh, bones and nerves. Disease is not a social role, but it makes sense to consider 'the sick role' as a social position with norms of appropriate behaviour (Parsons, 1951). It follows that it is plausible to argue that I have an illness, but also that I 'do', or perform my illness. The value of the phenomenological critique of Cartesianism is to demonstrate that consciousness is embodiment and also intentional.

For sociology, the limitations of phenomenology are determined by its exclusive focus on *my* body and the body of the *other*. The body is always socially formed and located. What it is to be man or woman is a social definition, since physiology is always mediated by culture. As Foucault (1980b) has shown, having a 'true' sex is the outcome of medical/cultural practices which in the case of hermaphrodites preclude the possibility of maintaining two sexual identities. 'Freaks' are also socially constructed (Howell and Ford, 1980).

While it is true phenomenologically that we have rulership over our bodies, it is never true socially in the sense that the social reproduction of populations is subject to institutional regulation ('the incest taboo'), power (in the form of patriarchy), ideology (in the contrast between desire and reason) and economics (the requirement of stable distribution of property through the household, typically in the form of unigeniture). Societies have traditionally been organized under the combination of gerontocracy/patriarchy in which the sexual behaviour of subordinates is regulated by the phalanx of God, king, priest and husband. The problem of the body is thus not simply an issue in epistemology and phenomenology, but a theoretical location for debates about power, ideology and economics.

# 3

# The Body and Religion

Man is a creation of desire, not a creation of need.

Gaston Bachelard, *The Psychoanalysis of Fire*

## Capitalism, Desire, Rationality

At various times in the history of sociology, the analysis of religion has been
seen as central to the theoretical development of the basic conceptual concerns of social science. The problem of religion was the starting point of the
young Marx's critique of social relations; it provided the focal point of the
sociology of Durkheim and Weber; the Paretian framework of nonrational
religious beliefs provided the groundwork of the Parsonian voluntaristic
action theory. At other times, the analysis of Christian institutions in secular society has appeared to be a relatively trivial branch of organization
theory and, as such, marginal to the perennial issues of sociology as a critical discipline. In response to this pendulum-swing development of the
sociology of religion, Berger and Luckmann (1963) proposed that, in order
to be grasped as central to the sociological enterprise, religion had to be
treated as a crucial dimension in the legitimation of everyday activities and
thereby constitutive of our 'knowledge' of social reality. The sociology of
religion was thus allied with the sociology of knowledge or, more precisely,
it was elided by the sociology of knowledge since religious beliefs became
merely a sub-category of social 'knowledge'. In this sense, the sociology of
religion could only remain influential by translation into a more general
analytical framework. Alternative theoretical strategies might involve not so
much the elision of the sociology of religion, but its intrusion into other sociological sub-disciplines for the purpose of contrast and opposition. My
intention is to suggest some important, but largely neglected, theoretical
connections between the sociology of religion and medical sociology (Turner,
1983). These connections could be elucidated by, for example, contrasting
the difficulties in the conceptualization of 'religion' and 'health'. Instead of
approaching this issue at the level of conceptual and methodological questions, this study insinuates the theoretical questions by an examination of the
historical contradictions between theology and medical practice in order to
introduce an additional dimension to the debate about the secularization of
religion in Western civilization. My thesis is, quite simply, that the generative
question in sociology of religion and medical sociology is the problem of the
body in society.

The notion that the institutionalization of anatomy is a major problem of social order is in fact implicit, not in Berger and Luckmann's article on knowledge, but in their influential study of social construction (1967). Following the work of Arnold Gehlen (Berger and Kellner, 1965) on biology and institutions, Berger and Luckmann argue that, although the biological constitution of men presents a set of physiological limitations on social relationships, the central feature of human biology is one of pliability and plasticity. For example, although human beings are equipped with sexual drives, these biologically rooted needs may find their outlet in a variety of institutions and practices: monogamy and promiscuity, heterosexuality and homosexuality, rape and prostitution. The implication is that the variability of human sexuality has to be channelled into certain socially constructed, routine patterns if social stability is to be maintained. Their theory of socially constructed human universes is thereby based on the argument that there are necessary tensions between body and society, and between self and body. The potential anarchy of biological satisfaction must be subordinated by a variety of institutional controls – especially the family – if the nomos of interpersonal interactions is to be preserved. At the level of personal experience, there is also an unstable relationship between subjective and objective worlds in that human beings can be said to have and to be bodies. In raising these questions, Berger and Luckmann not only suggest important areas of common concern between sociology and philosophy (such as the mind and body problem), they also link the sociology of religion with those issues which have been constitutive of sociology and psychoanalysis.

The argument that there is a fundamental incompatibility between the satisfaction of human instinctual needs and the requirements of civilization has occupied a prominent place in Western social philosophy. While in the original Hobbesian formulation of social contract theory human existence outside society was 'nasty, brutish and short', Rousseau's doctrine that social life is artificial and inimical to personal moral autonomy has probably been more influential in contemporary sociology. In Durkheim's analysis of the anomic division of labour, without adequate social restraints the instability of human desire represents a powerful suicidal tendency in human personality. A similar opposition between desire and social order was the underlying postulate of Malthusian demography, but the most pessimistic formulation of the contradiction between happiness and civilization was elaborated in Freudian metapsychoanalysis. For Freud, civilization is bought at the cost of instinctual satisfaction by the imposition of the demands of the superego over the id through the internal mechanism of guilt and conscience and the external mechanisms of social control. Civilization requires discontent. Unfortunately, the incorporation of Freudian psychoanalysis into sociology has often, as in Parsons's *The Social System* (1951), resulted in an over-integrated view of the relationship between the organism and the personality system – an incorporation which has, however, been noted by critics of functionalism (Wrong, 1961).

How does this traditional analysis of the contradiction between desire and social control relate to the sociology of religion? In rather simple terms, it may be suggested that, whereas early social contract theory explored the contradictions between personal happiness and social order, sociology of religion has centred around the apparent incompatibility between meaning and civilization. In Weber's two addresses on vocation in science and politics, one finds the classic statement of how the abandonment of the garden of enchantment prepares the way for a society organized around the principles of rational calculability and means–end rationality. However, scientific stability and predictability are bought at the cost of meaning, since, given Hume's naturalistic fallacy, normative statements can never be deduced from statements of fact. From the point of view of moral coherence, every advance in scientific understanding entails a reduction in the meaningfulness of the social world. Indeed, the quest for abstract knowledge beyond immediate technical needs must itself be irrational as a calling which can secure no normative legitimation. Although Weber's sociology of religion is conventionally approached in terms of a contradiction between meaning and knowledge, there is a major component of Weber's analysis of religions as various systematizations of irrational salvational paths where the opposition between body and meaning becomes critically important (Turner, 1981). Despite Weber's personal disapproval of Freudianism, there are important theoretical parallels between the attitudes of Freud and Weber to religion. In Weber's sociology, there are two specific points at which we find the argument that the advances of civilization require the suppression of orgiastic antinomian satisfactions.

The central theme of Weber's *The Sociology of Religion* (1966) is the question of how various religions, in formulating the relationship between the sacred and the profane, develop salvational paths of asceticism or mysticism in response to the 'world'. These formulations occupy a continuum between world acceptance and world rejection. While in his analysis of Christianity as a religion of brotherly love Weber is primarily concerned with religious beliefs in relation to wealth and power, the problem of the sexuality of the human body plays an important part in his conceptualization of religious rationalization. Weber recognizes that religious behaviour originates in mundane interests in health and wealth – 'The most elementary forms of behaviour motivated by religious or magical factors are oriented to *this* world.' However, the routinization of charisma arising out of ecstasy, the intellectual systematization of religious beliefs under the influence of the priesthood, and the rejection of magical means to mundane ends by prophecy replace these mundane interests by 'otherworldly non-economic goals'. These mundane interests are, however, never entirely suppressed by the ethical systems of the official orthodoxy and there remains a permanent tension between the religion of the virtuosi and mass religion in which the laity turn to the cult of saints for prosperity and healing. Similarly, the orgy of primitive religious expression is either repressed or sublimated in more acceptable institutional forms. Even the 'specifically anti-erotic religions' still 'represent substitute satisfactions of sexually conditioned psychological needs' (Weber, 1966, p. 237). The irrational sexual impulses of the body constituted a special problem within the Christian opposition to the world, since the unpredictable nature of erotic desire represented a direct challenge to any rationally organized system of ascetic self-control. Opposition to sexual irrationality was not, however, an issue peculiar to Christianity since 'no authentic religion of salvation had in principle any other point of view' (p. 241). The rational control of these impulses through the institutions of celibacy and monogamy represents an important dimension of Weber's master concept of rationalization in which the emergence of labour discipline and asceticism in capitalism constituted a major historical turning point in the control of the body.

The second illustration of Weber's commentary on the tensions between instinctual life and social organization occurs in *The Protestant Ethic and the Spirit of Capitalism*. The Protestant Ethic thesis may be briefly summarized as follows: 'A man does not "by nature" wish to earn more and more money, but simply to live as he is accustomed to live and to earn as much as is necessary for that purpose' (Weber, 1965, p. 60). The conditions by which natural man is forced out of a situation of mere reproduction for the sake of immediate physical needs are two-fold, namely the separation of the worker from the means of production and the emergence of the doctrine of work as a calling. The worker cannot be enticed out of this simple enjoyment of use-values merely by an increase in wages since 'Labour must, on the contrary, be performed as if it were an absolute end in itself, a calling. But such an attitude is by no means a product of nature' (p. 62). In Weber's theory of capitalist rationalism, every advance of capitalist production, therefore, requires the subordination of immediate instinctual gratification, the disciplining of the body and the quest for an economic surplus which far exceeds the present needs of utility and simple reproduction. Capitalist production requires both abstinence from immediate consumption on the part of entrepreneurs, and sobriety and self-control by the labour force. In this sense, capitalist accumulation is distinctively irrational from the point of view of 'nature', of basic instinctual gratification.

Although in the discussion of religious abstinence Weber was more concerned with the problem of wealth, he also recognized that sexual abstinence was a significant component of Puritanism and capitalism. The answer to religious doubt and sexual temptation was the same – 'a moderate vegetable diet and cold baths' (p. 159). In this respect, the 'hygienically oriented utilitarianism' of Benjamin Franklin had converged with the ethical outlook of the 'modern physicians' who advocated moderation in sexual intercourse in the interests of good health (p. 263). The control of the body, like the rational control of investment and production, appears to be an illustration of the rational spirit of capitalism and, in general, a feature of Western rationalization, but 'by nature' such developments are highly irrational. If we link this perspective with the earlier discussion in *The Sociology of Religion* we can say that, for Weber, every extension of systematic, formal rationality in religion (such as the emergence of other-worldly, ethical monotheism) entails 'a characteristic recession of the original, practical and calculating rationalism' (Weber, 1966, p. 27) which is geared to the satisfaction of the biological needs of natural man.

It is well known that Weber's own attitude towards rational capitalism was typically ambiguous. On the one hand, the rationality of capitalism and modern science had liberated men from the garden of enchantment in which magic and primitive gods had ruled their lives. On the other hand, this rational enlightenment had left men without any intrinsic meaning for life since scientific reasoning can by itself provide no answers to the central moral questions of human existence. Capitalism is the 'iron cage' in which people are mere 'cogs'. With the collapse of a religious calling to labour, the vocation for wealth and knowledge beyond immediate, utilitarian ends and beyond the basic needs of physical reproduction had, in fact, taken on a distinctively irrational quality. There is consequently a strong argument for drawing a parallel between Freud's pessimistic *Civilization and its Discontents* and Weber's sociology of religion. In both perspectives, the growth of an industrial civilization required a primitive accumulation of wealth which had been acquired at the cost of instinctual gratification and 'primitive rationalism'.

In these introductory comments, it has been suggested that the problematic relationship between body and society is a basic, but often implicit, issue in classical sociology, social philosophy and psychoanalysis. The plasticity of human instincts requires a set of powerful social and cultural mechanisms in order to maintain social stability. Weber, Freud, Berger and Luckmann have regarded religious asceticism as a major element in the development of the civilized institutionalization of instincts (Elias, 1978). This prefatory exegesis could be extended to examine the ambiguous attitude in Marxism to the sociological importance of asceticism (Seguy, 1977). The point of these observations is not, however, to provide yet another history of the sociology of religion, but to suggest that the theoretical linkage between sociology of religion and medical sociology is internal and necessary rather than superficial and contingent. Both sub-disciplines are fundamentally concerned with the complex interface between cultural and physiological structures. While certain anthropological studies of the pollution theory of disease and historical studies of mental health have recognized the importance of the relationship between religion and medicine, the importance of this issue has been generally neglected in mainstream sociology. Although this relationship could be explicated at an abstract level, it is suggested here that the religion/medicine issue might initially be explored by an examination of the ambiguities of the Judaeo-Christian tradition with regard to health, disease and treatment.

### Sickness, Salvation and Medicine

In evolutionary theories of religion, it was typically assumed that Christian spirituality and moral teaching could be neatly distinguished from primitive religions which failed to make a necessary distinction between physical pollution and moral sinfulness (Steiner, 1956). Sin could not be expunged by magic, ritual, or other institutional means since Christianity had spiritualized primitive notions of physical evil. The scriptural evidence of this uniquely ethical view of conduct was that men are not defined by what goes into their bodies, but only by their intentions. William Robertson Smith went further by suggesting that, within Christianity, this ethical stance was most fully developed in Calvinistic Protestantism, while Catholic sacramentalism had partly retained the notion that the cure of souls and bodies could be effected through the rituals of penance. There is some historical justification for this interpretation, since confessional practices in Christianity have characteristically been based on an analogy between priest and physician (Hepworth and Turner, 1982). In the early thirteenth century, for example, the parallelism between health and salvation was a common theme of confessional manuals. Robert Grosseteste drew a powerful analogy between the sinner/confessor and patient/physician relationship, and between moral vices and physical sins. In addition to confession, shrines and pilgrimages were traditionally associated with physical cures. Cure of disease was sought through religious relics for which shrines were erected (Bonser, 1963). While it was recognized that secular physicians had a definite sphere of competence, certain diseases and illnesses were the special province of priests. The miracles of Jesus provided a strong theological warrant for the priest as exorcist of demons. Given this tradition in Christianity of ritualized healing, it is not surprising that, when transplanted to other cultural areas, much of the therapeutic ritualism of the church proved especially attractive to indigenous populations. This is particularly true of the social significance of patron saints of disease for the aboriginal peoples of the Spanish American colonies (Guerra, 1969).

It could be argued, as Weber does, that these practices were merely concessions to the popular demand for healing services which ran counter to the central orthodox tradition of spiritual Christianity. With few exceptions, all religions have been compelled 'to reintroduce cults of saints, heroes or functional gods in order to accommodate themselves to the needs of the masses' (Weber, 1966, p. 103). Against this interpretation, it can be argued that the care of the sick is an act of charity *par excellence*, that physical exposure to disease is indicative of heroic virtue, and that diseases which are the consequences of immorality find their compensation in the institutionalized grace of the Church. In Catholicism, Thomas Aquinas's *Summa Theologica* clearly established the care of the sick among the principal acts of Christian charity, while suffering, nobly borne, was a sign of internal grace as was, for example, exhibited in the life of Teresa of Lisieux. The theological interpretations of the meaning of the body and disease were, in fact, far more complex than the doctrine of charity and virtuous suffering would suggest. On the one hand, the body ('the flesh') is deep-rooted in Christian symbolism of sinfulness. The frailty and eventual decay of the human body and the inevitable physical finitude of human beings provided an obvious metaphor for original sin and natural depravity (Frye, 1954). On the other hand, disease could be regarded as a sign of divine election by which the righteous are purified and perfected by the trials of pain (Job and Lazarus provided the biblical figures for this motif). There were, thus, three crucial types of illness – leprosy, hysteria and epilepsy – which were traditionally regarded as both an outcome of human sinfulness and as a sign of sacred election. Of these 'sacred diseases', leprosy is particularly instructive in the context of religious definitions of illness.

In the mediaeval ecclesiastical tradition, leprosy was not clearly distinguished from venereal disease and was specifically associated with sexual promiscuity. The outward decay of the leper was a sign of inner profanity. The leper constituted both a moral and physical threat to the community and had to be separated from the population by dramatic rituals and other legal means. The church's office at the seclusion of a leper did not differ in fundamentals from the office for the dead, since the *separatio leprosorum* defined the leper as a ritually dead person. In the service, the leper bends before a black cloth by the altar 'after the manner of a dead man' (Clay, 1909, p. 273). Once lepers had been ritually secluded, they were by law prevented from inheriting property, entering public places, touching wells without gloves or eating in company. Under these conditions, the leprosarium became a refuge. While these rituals of exclusion gave overt expression to the condemnation of leprosy as a punishment for carnal lust, leprosy was perceived in terms of two contradictory attitudes: 'the disease was the sickness both of the damned sinner and of one given special grace by God' (Brody, 1974, p. 101). The disease was paradoxically known as the 'sacred malady' and the disease of the soul. There was a clear need to remove the association of sin and leprosy after the First Crusade when soldiers, fighting a holy war in the hope of personal salvation, returned home with leprosy. The parable of Lazarus provided one means of regarding leprosy as a divine affliction which, through patience, led eventually to a divine reward. The contradiction was eventually removed by the disappearance of leprosy as a common disease in the middle of the fourteenth century. It was replaced by syphilis which again came to be regarded as a divine punishment for human sins.

In historical terms, there are a variety of different but related theological views of sickness. Within the spiritual scheme of human affairs, disease might have a number of different functions: 'To enhance the merits of the just through their patience, to safeguard virtue from pride, to correct the sinner, to proclaim God's glory through miraculous cures and, finally, as the beginning of eternal punishment as in the case of Herod' (Temkin, 1977, p. 54). Despite the apparent diversity of views, one important theme is that disease is not religiously neutral, but shot through with deeper spiritual significance. The natural processes of the body were not merely events within an external, natural world but 'could conceivably be interpreted as a corporeal manifestation of the relationship between God and man and this meant of man's spiritual and moral life' (p. 54). Such an interpretation raised two immediate difficulties. First, the notion of disease, as caused by moral imperfections for which individuals could be held responsible, conflicted with the idea that disease had a divine origin. The problem of disease was thus part of a deeper problem about evil, divine justice and hence theodicy. Secondly, the church was forced to adopt a conciliatory attitude to medicine which appeared to intervene in the divine mechanisms of disease by offering cure and therapy. To what extent could medical cures be reconciled with the idea of moral responsibility?

### Medical Ethics and the Medical Fee

In Chaucer's Prologue to the *Canterbury Tales*, we find that the richly clad doctor

> . . . was rather close as to expenses
> And kept the gold he won in pestilences.
> Gold stimulates the heart, or so we're told.
> He therefore had a special love of gold.

These lines neatly indicate the problem of medical practice for any moral system grounded in supernaturalist presuppositions. Although medical intervention appears to conflict directly with the notion of disease as a sign of religious fate or as a dimension of a divine scheme for human affairs, a theodicy of pain can coherently incorporate medical therapy as an act of charity. However, if the physician charges a payment for his services, then it is difficult to reconcile the medical fee with the concept of charitable virtue. In particular, as in Chaucer's Prologue, it appears as if the unsuccessful doctor's career is parasitic upon the misfortunes of the body. The main solution to the moral dilemma of medical remuneration was found in the development of a code of practice by which the community sought some control over the client–physician relationship. The growth of the notion of the 'ideal doctor' in Jewish (Margalith, 1957), Greek and Christian culture produced a body of deontological writing out of which the professional ethic of medicine eventually emerged.

Within the framework of Greek science and philosophy, medicine as a practical skill of wandering craftsmen was, to some extent, socially stigmatized on two accounts. First, it was regarded as a manual, not mental, exercise as the name *techne iatrike* signified (Kudlien, 1976). The practical aspect of the occupation was even more pronounced where medicine was combined with surgery which literally meant 'hand-work' (*kheirergo*). Secondly, the fact that these craftsmen charged a fee distinguished them from noble callings, like mathematics, which were not aimed at and did not involve financial reward. Various professional ideologies were developed in classical antiquity in order to legitimate medicine as a virtuous occupation. Elements of the Hippocratic corpus dealing with philanthropy were emphasized and attention was drawn to the noble descent of Hippocrates. If the physician had inherited wealth, he could afford to practise without payment, or at least remuneration would be less important. In Greek medical treatises, it was, therefore, suggested that medical students should be free-born citizens from wealthy families. In medical deontology, norms were established to regularize the interaction between doctor and patient in those cases where a fee was required. Payment should not be requested too early, the physician should not be too overtly concerned with personal gain and the poor should not be exploited. The ethical component of Hippocratic medicine was further developed by Galen of Pergamon who argued that the principal aim of medicine was not financial gain but service to the community in the restoration of health. Various medical tracts assumed an almost casuistic character in claiming that the real difference between ignoble occupations and altruistic medicine was that, in the case of medicine, the fee was accepted but not requested. It was the request for payment, not acceptance of reward, that made certain occupations dishonourable.

The reception of the classical conception of medicine as a virtuous activity, despite the vexed questions attending the medical fee, in early Christianity posed the problem of how to assimilate a secular practice within a religious framework. One consequence of this tension between the religious and the secular was that in the period from the fifth to the eleventh centuries writing on the code of medical practice was dichotomized around ethics and etiquette (MacKinney, 1952). While medical ethics emphasized the general moral qualities of the medical calling, etiquette dealt with the practical, secular problems of doctor–patient interaction. Etiquette was concerned with the doctor's bedside manner, dress, language and deportment. In the religious tradition, medical practice was to some degree fused with the charitable work of the monasteries which permitted a certain *rapprochement* between Christian idealism and Hippocratic philanthropy. Despite the alliance of the monastic norm of charitable healing with the classical view, the norms of everyday medical practice for physicians were essentially this-worldly. The increasing influence of the medical school at Saleno gave a decisive impetus to the growth of a secular, individualist code of medical practice organized around etiquette rather than ethics (Bullough, 1966).

The debate concerning the medical fee in classical and Christian deontology was in fact indicative of a profound division in medicine relating to basic questions of theory, practice and values. Although the Hippocratic corpus contained a number of separate theoretical traditions, Hippocratic medicine divided into alternative perspectives, namely empiricism and rationalism. In the empiricist tradition, medicine was regarded as having a minimal, modest role in assisting the body to overcome disease in a manner which was determined by the organism itself. The physician did not alter the course of the natural process of recovery; he assisted the natural sequence of illness, crisis and restoration rather than intervening to change the natural pattern of events. The drugs which the doctor may have prescribed were selected on the 'basis of "similarity" in that they stimulate the organism to continue along the path which it has already chosen' (Coulter, 1977, vol. 1, p. 89). Since the internal laws of the body were ultimately unknown and unanalysable, the physician had to pay particular attention to the symptoms of the disease which provided the essential clues for the choice of therapy. The empiricist tradition consequently regarded medical knowledge as the product of medical experience and as subordinate to medical skill. By contrast, the rationalist tradition treated the internal processes of the human body as analysable in terms of knowable causal mechanisms. Medical knowledge was abstract, coherent and deductive; it was not the outcome of medical practice, but, on the contrary, guided practice in terms of rational knowledge. Medicine did not help the natural processes of the disease crisis but intervened to change the disease through the use of 'contrary' medicine. The empiricist doctor at best was the assistant of *physis*; by medical intervention, the rationalist subordinated *physis* to theory.

The difference between these two approaches to medicine can be illustrated in terms of the humoral theory of disease. Although rationalist and empiricist positions in classical medicine were humoralist, empiricists argued that the body was constituted by an unknown, indefinite number of humours. Cure of the disease was to be effected by the coction of isolated humours through the normal processes of evacuation of sweat, urine and faeces. The doctor followed the course of the illness by careful examination of the body's excretions and secretions; the therapy of the empiricist physician typically involved a combination of 'diet, exercise, hot baths and applications (to promote coction), and laxatives (to promote evacuation)' (Coulter, 1977, vol. 1, p. 7). In the rationalist tradition of Aristotle and Galen, medical understanding of human pathology was founded on four specific humours: blood, phlegm, yellow bile and black bile. It was improper combinations of these humours which were the causes of specific illnesses. The humoral theory was elaborated to include four qualities (hot, cold, wet and dry), four elements (air, fire, water and earth), and four seasons (spring, summer, autumn and winter). The juxtaposition of these various qualities resulted in property–space diagrams in which a great diversity of diseases could be unambiguously located by deduction. The important difference between empiricism and rationalism was that the former treated each disease as a problem of the individual in the context of the patient's total life, while the latter dealt with classes of disease rather than individual cases. The goal of rationalism was, with the aid of a logically consistent theory, to penetrate behind the symptoms to the disease entity which was the cause of pathology.

Rationalism, particularly in the form of Galenic humoralism, became the dominant paradigm of medical practice until experimental anatomy in the Renaissance began to question much of the received wisdom of Greek medical science. The triumph of rationalism was closely linked to the professional needs of medical practitioners. Empiricism was the perspective of medical practitioners who were not differentiated from other craftsmen and whose knowledge of disease was not different in kind from that of laymen. Rationalist humoralism was not commonsense knowledge based on experience and claimed to comprehend hidden causes of illness which could not, therefore, be observed and treated by the ordinary layman. The more medical theory could be removed from direct observation of symptoms and everyday experience, the greater the inequality of social status between doctor and patient. Given their belief in the specificity of disease in each individual patient, the empiricist physician had to make painstaking and lengthy observations on the patient before the appropriate therapy could be determined. Rationalism, by contrast, provided apparently quick, sure answers and a therapy which was determined by knowledge of cases, not individual patients. Finally, the test of empiricist therapy was whether the patient recovered and this situation meant that the individual doctor was particularly exposed to criticism from patients and relatives when the cure was ineffective. For the rationalist physician, humoralism was a valid theory; if the cure appeared to fail, it was because the patient had not followed the medical regimen strictly. The patient, not the theory, was responsible for lack of curative success.

## Medicine as a Secular Practice

There is a common assumption in sociology that the transition from feudalism to capitalism in the seventeenth and eighteenth centuries in Europe represented *the* great transformation which divided history into its traditional and modern phase. Although there is widespread disagreement as to details, it is generally held that traditional society was rural, stable and religious, while modern society is urban, complex and secular. It is against the theoretical background of the great divide doctrine that the study of medical history is of particular interest to the sociologist of religion. In Western society, the dominant medical paradigm has been, at least from the fifth century BC, pagan and this-worldly. While there has been some fusion of the Greek conception of philanthropy with the theology of Christian charity, the Church found itself forced to compromise with Greek medicine. As we shall see, Christian views of medicine and medical practice have often been an oppositional force against the mainstream of medical secularity. Thus, in medical history, the transition to capitalism did not affect the essentially this-worldly professionalism which was part of the Greek legacy from antiquity.

The rationalist character of Greek medicine may itself be an effect of early professionalization in a competitive market context. The status-conscious, secular physician looked to medical science as a basis for distinguishing his practice from the popular, magical remedies of leechcraft. This professional hostility to superstition and popular belief may have been the basis of the overt secularity of the Hippocratic treatise on *The Sacred Disease* in which it was denied that epilepsy

> is any more divine or sacred than any other disease but, on the contrary, has specific characteristics and a definite cause. Nevertheless, because it is completely

different from other diseases, it has been regarded as a divine visitation by those who, being only human, view it with ignorance and astonishment. (Lloyd, 1978, p. 237)

In denying the sacred nature of epilepsy, the Hippocratic collection attacked those charlatans who made a profit from popular ignorance and superstition. An alternative, naturalistic explanation for epilepsy was proposed, namely a discharge of superfluous phlegm into the blood vessels of the brain. While such explanations of disease appear to be unambiguously secular, the interpretation of Hippocratic medicine is in fact more complex, since the Hippocratic corpus equated nature with divinity. One of the instigating circumstances of epilepsy was thought to be a southerly wind. Since the wind and other forces of nature are divine 'there is no need to regard this disease as more divine than any other' (Lloyd, 1978, p. 251). This 'rationalistic supernaturalism' (Edelstein, 1937) combined with the professional requirement of secular medicine to drive out any association between leechcraft and Hippocratic medicine. It was also for this reason that the association of Greek medicine with the cult of Asclepius did not influence the secularism of medical practice. Fictitious claims of descent from the hero-god, Asclepius, provided a sense of occupational communalism for itinerant medical craftsmen without influencing the secular nature of their craft (Sigerist, 1961).

The dominant Greek medical tradition was, therefore, secular and pagan with a pronounced theoretical emphasis which involved a definitely negative attitude towards medicine as a practical craft. Within the profession of medicine, there was a sharp differentiation between the theoretically oriented physician and the mercenary healer of bodies, the medicus. In addition, surgeons were merely manual workers who, in Britain for example, were not institutionally separated from barbers until the early seventeenth century. The separation of theoretical medicine from clinical experience was further intensified by the growth of university-based medical schools. The development of Salerno, Bologna and Paris as medical universities caused control of medicine rapidly to come under the monopoly of university-trained physicians. Various exclusionary practices, such as the importance of Latin as the principal vehicle of medical education, gave the profession greater control over entrance to medicine. The secular idealism of the Hippocratic tradition became increasingly important in controlling the activities of physicians and in justifying the exclusionary devices of the profession. Thus, 'To sell the public on the exclusionary tactics of the university trained practitioner, the need for regulation had to be couched in terms of ideals, but to gain the support of the would be professional it was necessary to emphasize self interest' (Bullough, 1966, p. 93). The legitimacy of the medical fee for a secular profession was thus an aspect of increasing institutional specialization of medicine in conjunction with humanitarian idealism. The emergence of a secular profession, organized around the pagan humanism of Aristotle and Galen, to a dominant position within mediaeval and Renaissance universities created a number of important tensions with the dominant religious culture of Catholic Europe and consequently opposition to secular medicine often took a decidedly Christian character.

While medical theory conceptualized the human body by reference to mechanistic analogies in which disease was the effect of known, physical causes, the Christian critique of Galenic medicine typically emphasized the importance of the patient as an individual uniquely related to the natural and social environment. This religious perspective was particularly important in the medical theories of Paracelsus (1493–1541), van Helmont (1578–1644) and Samuel Hahnemann (1755–1843). Doctors within the Paracelsan tradition were highly critical of physicians who treated the disease rather than the patient by attempting to oppose the natural process of recovery by 'contrary' medicine, who adhered to pagan medical theories divorced from clinical experience, and who had a materialistic and this-worldly conception of their professional calling. Christian opposition to secular medicine tended to embrace the empiricist critique of rationalism, infusing humanist empiricism with a religious conception of medicine as a calling. Paracelsus in particular was conscious of the stark incompatibility between the Christian ethic of love and the humanist medical tradition which he regarded as a thin disguise for acquisitiveness. Paracelsus's aim was to 'oust the false, pagan, Galenic and Arabic teachings of the schools' (Coulter, 1977, vol. 1, p. 356) with the Christian doctrine of love, and his religious beliefs constituted not only a radical change in the doctor–patient relationship, but also a view of medical practice which was minimalist rather than interventionist. The least medicine was always the best medicine since recovery from illness was ultimately part of God's will and subject to the natural healing processes of the body.

The justification for medicine based on charity and practical experience involved more than simply an appeal to the doctrine of brotherly love. To discover through medical practice the secret of the inner processes of the body was also to discover the hidden purpose of God. Paracelsus thus drew a parallel between faith as knowledge of what is hidden to natural man and medicine as knowledge of processes which are invisible to the eye. It can be argued that this emphasis on experience as the key to the knowledge of God and nature provided an important link between Paracelsan medicine, Baconian science, the scientific revolution of the seventeenth century and the homeopathic medicine of Hahnemann (Debus, 1972). While this argument in general supports my claim that the secularity of medical knowledge and practice have traditionally presented acute problems for Christianity, there are a number of specific difficulties with this thesis as presented in Harris Coulter's *Divided Legacy* (1977). There are, for example, rather obvious problems of interpretation with regard to the medical theories of Paracelsus, who remains a particularly elusive figure in medical history (Temkin, 1952). More substantively, Coulter attempted to connect Christian opposition to Galenic medicine with Neo-platonism and mysticism. One implication of this interpretation is that ascetic Christianity may have had a greater affinity with rationalist medicine, but Coulter fails to explore this particular implication. Any medical tradition, whether empiricist or rationalist, which provides a medical regimen of disciplined behaviour – diet, exercise and sobriety – is likely to prove attractive to religious asceticism. Despite the secularity of medical professionalism, there may be important affinities between those forms of personal denial which are implicit in the medical regimen and an ascetic life-style.

## Capitalism and the Body

On crude Marxist grounds, it could be argued that capitalists have a contradictory attitude towards the health of the labour force. The competition between capitalists forces down wages and the mechanization of production creates an industrial reserve army of unemployed workers. The effects of these two features of competitive capitalism on the standard of health of the British working class were graphically portrayed by Engels (1952) in *The Condition of the Working Class in England in 1844*. By contrast, it is also argued by Marxists that capitalists require an efficient, disciplined and sober labour force in order to maintain continuous production and maximum utilization of machinery. While Weber noted the importance of asceticism for the bourgeoisie, it has also been suggested that the Protestant sects were 'functional' for capitalism in creating a healthy, hard-working and sober working class. If mystical, Neo-platonic Christianity provided a doctrine which opposed secular, Galenic medicine, there is, by contrast, an obvious relationship between Christian asceticism and medical theories which stressed the importance of restraint for physical health. There is, thus, a *prima facie* parallel between the idea of the medical regimen and religious rules of ascetic discipline, in that both are addressed to the government of the body. In England, the conjunction between Wesley's spirituality and medical asceticism provides an important illustration but this grew out of a background which was influenced by rationalist, not empirical, medical theory.

In the late seventeenth and early eighteenth century, medical theory, partly through Descartes' rationalist mind–body dualism, came to be markedly influenced by mathematical and chemical models of the body, which was conceived as a complex machine. A variety of iatromathematical and iatromechanical theories of disease gained in popularity under the influence of Boerhaave's school of medicine at Leyden (Underwood, 1977). In Britain, the dissemination of these mathematical models of the body owed a great deal to the influence of the Aberdonian doctor George Cheyne (1671–1743). Following Harvey's discovery of the circulation of the blood, Cheyne argued that the body was merely a series of canals and that pathology could be properly studied mathematically. While Cheyne anticipated the rapid development of Newtonian medicine, he is best remembered as an advocate of diet, moderation in drink, and light exercise for healthy living and mental stability.

Cheyne's principal publications were devoted to advertising the benefits of moderate asceticism for mental stability and physical well-being. In addition to his writing on natural religion, his main medical treatises were *Essay on Health and Long Life* (1724), *The English Malady* (1773), *Essay on Regimen* (1740) and *Natural Method of Curing the Diseases of the Body* (1742). These works went through many editions in his own lifetime and through his clients and friends, who included Samuel Johnson, John Wesley, David Hume, Alexander Pope, Samuel Richardson and the Countess of Huntingdon, Cheyne enjoyed considerable popularity and influence. Wesley was particularly impressed by Cheyne's medical asceticism. It was from reading Cheyne that Wesley learnt, according to his *Journal*, 'to eat sparingly and drink water' and much of Cheyne's practical advice found its way into Wesley's popular *Primitive Physick or an Easy and Natural Method of Curing Most Diseases* (1752). Although Wesley's brand of asceticism was based on an eclectic interpretation of the Christian tradition, it is clear that Cheyne's view of the benefits of modern living was influential in the standards which were set by Wesley for the new Methodist societies. Cheyne's regimen had initially been aimed at an elite who were suffering from excessive consumption of meat and wine, but, through Wesley's *Primitive Physick*, these medical guidelines became part of the Methodist discipline for a much wider section of the community. In a period of expanding industrial production, the government of the body became, not only a sign of social standing, but an outward indicator of spiritual virtue. Weber, in emphasizing the importance of emotion and conversion in early Methodism, neglected this contribution of Wesleyanism to the rational conduct of the middle and working classes. If there is, at a general level, a set of tensions between the Christian doctrine of love and the medical fee as the financial basis of a secular profession, in the eighteenth century pietist asceticism merged with the medical regimen of healthy living to produce a moral code which was compatible with the capitalist's interest in a disciplined work force. The duty to be healthy became part of a calling to world mastery and self-control.

There were, however, additional and perhaps more urgent reasons for capitalists to take an active interest in standards of health in the late eighteenth and nineteenth centuries, namely that urban squalor represented a health hazard to all social classes because of the epidemics which were common in the new industrial cities. In this period, pietism provided a significant moral component for the sanitation movement in both Europe and North America. From a religious perspective, disease is redolent with moral implications. The epidemics which accompanied nineteenth-century urbanization were seen to be the effects of human irresponsibility and, in particular, as consequences of human mismanagement of the social environment. Disease was part of the disequilibrium between body and environment, resulting from abuses of diet, poor hygiene, immorality and, above all, filth.

Health was the manifestation of the dialectic between order and chaos, purity and danger, responsibility and immorality. For pietists, the movement for urban sanitation was symbolic of human responsibility towards the physical environment in removing the evils of poor housing, congested sewers and inadequate ventilation. For religious reformers like Florence Nightingale, the order and cleanliness of the hospital was a microcosm of that harmony which was the ideal equilibrium of the larger society. Human culpability for infection consequent upon bad management and negligence brought volition and responsibility to the centre of the aetiology of disease. The concept of social pollution as a congenial environment in which diseases underwent a process of fermentation was especially important to pietist sanitarianism. The notions of 'fermentation' and 'purefaction' became heavily charged with moral significance and, for Nightingale, the explanation of

> [d]isease causation in terms of fermentation was effective in emphasizing the purity of God's atmosphere and the necessarily culpable role of man in polluting it. As compelling justification for the social activism demanded by her personality, the image of zymotic disease could hardly have been more appropriate. (Rosenburg, 1979, p. 123)

With their commitment to moral responsibility and an environmentalist theory of disease, pietists were opposed to the concept of specificity of disease as the product of definite germs which appeared to attack populations on a random, moral basis. Pietist sanitarians like Edwin Chadwick and Dr Southwood Smith were also hostile or indifferent to new developments in immunology and bacteriology, because germ theory could not be easily amalgamated with a moral doctrine of disease as a consequence of human irresponsibility.

There was a Rousseau-like quality to religiously motivated opposition to the specificity of disease, immunization and vivisection in the middle of the nineteenth century since, amongst pietists, it was fashionable to believe that civilization breeds social and physical corruption. The answer to the diseases of civilization had to be sought in salvation and adequate sewage, without which technical advances in medicine would be ineffectual. Some anti-vivisectionists went even further to argue that experimental medicine displaced human responsibility for sin and disease onto the animal world. The best antidote to disease was healthy living, personal cleanliness and a pure water supply. The theme of social and physical purity also produced a marked hostility to vaccination, because the preventive medicine of Pasteur involved the introduction of 'impurities' into the body.

In the late nineteenth century, the pollution theory of disease was harnessed to the evolutionism of Darwin and Spencer. The moral evolution of the human species would render vivisection and vaccination unnecessary, but the need for universal sanitation required increasing state intervention and this political requirement tended to contradict the essential individualism of pietism. Contagionism implied that, as it were, diseases select their victims at random, regardless of their moral standing and virtuous existence because the healthy might be infected despite their morally virtuous existence. Pietists tended to be anti-contagionists on moral grounds, but also because quarantine measures involved state control of the individual (Ackerknecht, 1948). However, religious sanitarians could not avoid the problem that good housing, pure water and efficient sewerage systems require legislation and the intervention of the state into the private domain of the individual. In Britain, Spencerian evolutionism and pietist sanitarianism paradoxically involved state legislation to create a physical environment for moral regeneration. The same paradox confronted liberal sanitarians in the United States where religious activism was the initial driving force behind the movement for public health. In France, by contrast, the public hygiene movement was dominated by men who had been trained in Napoleon's armies, and whose social outlook had been shaped by the legacy of Saint-Simon's *physique sociale* and the 'medical anthropology' of Cabanis. Despite the variety of religious and philosophical influences on the people who were the spearhead of social sanitation, public health involved state control of the urban environment.

The movement to improve the physical and moral condition of the 'dependent classes' in competitive capitalism thus provides us with an interesting Weberian parable. The religious 'world images' which had been the ideological force behind public health reforms had the unintended consequence of fostering secular, bureaucratic control of the environment. The ethic of healthy living could be said to have an affinity with the spirit of bureaucratic medicine, but, once constructed, bureaucratic sanitation no longer required the prop of religious asceticism. Individual choices about health are no longer set within a religious framework linking sin to disease and these personal choices are also set within the limits of legislation and secular campaigns against alcoholism, smoking and obesity. Such campaigns have to presuppose that 'life is worth living', but modern medicine can provide no basis for such a presupposition. Although modern medicine is forced to deal with decisions of life and death, it cannot contribute to the question of

> [w]hether life is worth while living and when – this question is not asked by medicine. Natural science gives us an answer to the question of what we must do if we wish to master life technically. It leaves quite aside, or assumes for its purposes, whether we should and do wish to master life technically and whether it ultimately makes sense to do so. (Weber, 1961, p. 144)

Scientific medicine cannot address and does not provide answers to those problems of pain and suffering at which traditional theodicy was directed. The Weberian parable is thus completed by the fact it is precisely in those areas of social relationships where intimacy and meaning (Skultans, 1974) are at a premium that certain forms of 'religious medicine' – faith healing, spiritualism, exorcism – continue to flourish as an alternative, and frequently in opposition, to technically sophisticated, secular medicine.

My argument has been that there are important, but neglected, empirical relationships between religion and medicine, and equally significant interconnections between the sociology of religion and medical sociology. Insofar as pain, suffering and finitude constitute the principal referent for all theodicies, the body and meaning are critical concerns for all salvation religions. It was in this way that the problem of the body entered into tension between religion and the world in Weber's comparative sociology of salvational systems. It is also one of the perennial topics of social philosophy. The sublimation of instinctual life was a major feature of the philosophy of Nietzsche and the psychoanalysis of Freud, while the question of reproduction permeated, not only Malthusian demography and Social Darwinism, but the whole political economy of the eighteenth and nineteenth centuries.

To raise the question of the relationship between sociology of religion and medical sociology is thus not to introduce an extraneous side issue, but to return sociological discussion to a set of problems which are constitutive of the discipline. These problems have in this argument been illustrated by a series of case studies: Christian and Greek attitudes towards the medical fee; the secular content of medical ethics; the association between diet and asceticism; and the pietist background to sanitarian reforms. These studies illustrate both the perennial tensions between religion and the profane body, between Christian charity and secular medicine, but also the periodic alliances between the medical regimen and asceticism. Christianity arrived at a number of institutional solutions for controlling the body (such as limited sexuality within marriage) and a variety of compromises with secular medicine (such as an ethical vocation in medicine). Despite these compromises, the Christian churches have traditionally found themselves in some state of opposition to secular medicine. The medical fee was seen to be incompatible with the Christian doctrine of disinterested charity towards the poor and the sick. In addition, the explanation of disease by reference to specific cases of a chemical or physical character without reference to divine intervention conflicted with the religious notion of illness as a moral event for which the individual was accountable.

In making such sharp contrasts between body and meaning, the question of psychosomatic phenomena, where issues of meaning and intention remain important, has so far been generally ignored. While the relationship between religion and madness is historically interesting, it does not raise issues which are qualitatively different from that of religion and physical health. Just as secular medicine viewed the body as a machine, so the mechanist–vitalist debate about the nature of the mind was resolved in favour of biology and physiology. The general trend of rationalist medicine is to treat 'deviant behaviour' (madness, alcoholism, drunkenness, homosexuality, and so on) as, in fact, a problem of chemistry. As with mechanistic theories of diseases of the body, psychosomatic medicine has become, in the last analysis, indifferent to those questions of meaning and morality which have been central to all religious world-views. From this sociological perspective, it would appear that one could perceive secularization as that process by which other-worldly views of disease were gradually replaced by amoral, mechanistic theories of the body under the impact of an increasingly powerful medical profession.

## Conclusion

The sociology of religion appears to be particularly prone to theoretical crisis and self-criticism. The secularization of religion in modern societies threatens to rob the sub-discipline of empirical content in that the sociology of religion would then concern itself with obsolete beliefs and practices or, at best, become a minor feature of a more inclusive sociology of beliefs. While such academic pessimism may be well founded, this chapter has attempted to outline a programme of restoration of the sociology of religion to a place of theoretical importance by suggesting an alliance with medical sociology. Following Durkheim, the sociology of religion may be construed as the analysis of the social interaction between the sacred and the profane. Historically, the most potent symbol of the profane world is the human body. The body is dangerous and its secretions, particularly semen and menstrual blood, have to be enclosed by ritual and taboo to protect the social order. Yet, at the same time, the body is sacred. The charisma of holy individuals typically flows through their physical secretions, being stored up in blood and sweat. The concept of 'salvation' itself is intimately bound up with the health of the body as the verbs 'to save' (the soul) and 'to salve' (the body) indicate. The mediaeval penitentials gave expression to the notion that human sinfulness was a conflict between the body and soul which found its resolution in the healing ministry of the sacraments under the institutionalized monopoly of the universal church. The sacrifice of Christ's body had produced a beneficial supply of healing charisma which the church stored up in a Treasury of Merit and which could be redistributed through confessional absolution and through the wine and bread of Holy Eucharist. In Christianity, the bread of communion was symbolic of spiritual and physical reconciliation and well-being. Comparatively, it is interesting to note that in Islamic Sufism, the *baraka* of the saints is also associated with bread which is symbolic of plentitude and health (Turner, 1974). The parallelism between spiritual and physical well-being is thus a common theme of the Abrahamic tradition.

Contemporary sociology of religion has generally neglected this intimate relationship between body and belief, between medicine and religion. With one or two exceptions, such as Norman O. Brown's *Love's Body* (1966) or Herbert Marcuse's *Eros and Civilization* (1969), the subtle linkages between biology and sociology have been widely neglected. Within the sociology of religion, this absence of theoretical reflection is somewhat surprising, given the ultimate location of the problem of meaning in the finitude of human existence (Turner, 1983). To some extent, this theoretical negligence is also characteristic of medical sociology, which, because of its practical importance for institutionalized medicine, has often devoted itself to limited empirical issues: doctor–patient interaction, hospital administration, social factors in the aetiology of sickness, the role conflicts of nurses, and so forth. Medical sociology has, however, also concerned itself with the meaning of illness and the subjective significance of pain. Both the sociology of religion and medical sociology converge on the question of human suffering and the indignity of death; they are therefore inevitably cultural responses to the problem of theodicy. The intellectual task of making that convergence theoretically systematic and deliberate is consequently not only an important item on the agenda of contemporary sociology of religion, but of the sociology enterprise as a whole.

# 4

# Bodily Order

## Hobbesian Materialism

It has been argued that the problem of order (namely the question 'how is society possible?') is fundamental to any social theory. The question has traditionally divided sociology into two distinctive branches of enquiry. Conflict theory argues that social order is deeply problematic and, insofar as it exists at all, is brought about by coercive circumstances, political constraint, legal force and the threat of violence. Consensus theory treats social conflict as abnormal by arguing that social stability is brought about by fundamental agreements over social values and norms which are instilled in social members by the process of socialization which rewards conformity to existing arrangements. This clear-cut analytical division rarely occurs in a 'pure' form, since social theories tend to adopt elements of both types of explanation. For example, the concept of 'hegemony', which is often used to explain the relative stability of capitalist societies, is a mixture of both cultural consensus and political coercion. The debate about social order in contemporary sociology owes a great deal to the formulation of the so-called Hobbesian problem of order in Parsons's *The Structure of Social Action* (1937).

Parsons's study of shared values as the ultimate bed-rock of social order was a reply to positivist theories of social action. For example, rational positivism argued that human action was to be explained in terms of the rational pursuit of egoistic interests such that any deviation from interest was irrational action. Hedonistic psychology and utilitarianism adopted similar views of the nature of human behaviour: human behaviour was rational in that human actions were aimed at the maximization of pleasure and the avoidance of pain. Parsons's argument was that such a model of human action was incapable of explaining social order and could not successfully distinguish between 'action' and 'behaviour'. If human beings rationally pursued their individual interests, they might quite rationally employ force and fraud to achieve their ends, but it would then be difficult to account for social order and stability in the widespread presence of force and fraud. In Parsons's argument, there has to be some minimal agreement about values in society for social relationships to exist at all. For example, there are certain agreements about avoiding fraud and such agreements make society possible. One criticism of social contract theory as expressed by utilitarianism is that it would not account for the binding nature of such contracts over self-interested parties. Parsons's second line of criticism was that, while behaviour might be explained within a behaviouristic framework, social action could not be. Action involves the choice of certain ends and the selection of means to achieve these ends in terms of shared standards or norms. Action is purposeful not simply in terms of a pain/pleasure principle but in terms of intention and choice. Instrumental rationality is not the only definition of rational action, since actions may be regarded as rational if they are in conformity with certain values, which are not themselves reducible to biology, environment, economic interests or psychology. Parsons's solution to the Hobbesian problem of order in terms of shared values was thus also intended to be an answer to the limitations of positivist epistemology (Hamilton, 1983).

Parsons's approach to the nature of social order has been itself the object of considerable criticism (M. Black, 1961; Dahrendorf, 1968; Gouldner, 1971; Rocher, 1974). A number of basic objections to Parsonian functionalism have arisen from this debate. One standard criticism of the Parsonian model is that it cannot provide an adequate theory of social change, because it exaggerates the level of value coherence within societies. Another criticism is that, while values may be normatively adhered to, general values may also be accepted pragmatically because the alternatives to these values are not available or inadequately perceived (Mann, 1970). Parsons's treatment of values also assumes that the same values are held by all members of a society; an alternative position argues that different classes in a society may hold different value systems and that coherence of societies is to be explained by economic constraints on action (Abercrombie, Hill and Turner, 1980). Another general criticism of Parsons is that his sociology is couched at such an abstract level that it does not lend itself to empirical falsification.

Although Parsons has provided much cogent counter-criticism (Parsons, 1977), and although his approach is still subject to on-going assessment (Alexander, 1982), it is interesting to reflect on the nature of the problem of order by returning to its formulation in the work of Hobbes. Parsons referred to Hobbes's solution to the problem of the possibility of society as 'almost a pure case of utilitarianism' (Parsons, 1937, p. 90) and yet it would be far more accurate to see Hobbesian philosophy as a pure case of materialism. Hobbes's aim was to reconstruct political philosophy in terms of scientific principles, because he regarded existing philosophy as underdeveloped or uncultivated: 'Philosophy seems to me to be amongst men now, in the same manner as corn and wine are said to have been in the world in ancient times. For from the beginning there were vines and ears of corn growing here and there in the fields; but no care was taken for the planting and sowing of them' (Molesworth, 1839, vol. 1, p. 1). Hobbes's starting point was the geometry of bodies and the principles of motion. His materialist philosophy was developed in three stages: the motions of bodies in space, the psychology of men and finally the analysis of such 'artificial' bodies as the corporation and the state. Thus he wrote that his intention was to discuss 'bodies natural; in the second, the dispositions and manners of men; and in the third of the civil duties of subjects' (vol. 1, p. 12).

Hobbes started out characteristically with a definition of body as extension and referred to man as an 'animated rational body'. Hobbes went on to argue that in nature men enjoy a general equality of their four main characteristics: strength of body, experiences, passion and reason. However, this equality is undermined by vanity, appetite and comparison; because men will seek to preserve their lives, they necessarily come into conflict with other men: 'For every man by natural necessity desireth his own good, to which this estate is contrary, wherein we suppose contention between men by nature equal, and able to destroy one another' (vol. 4, p. 85). While men in nature live in a state of war, they also have reason and it is reasonable for men to pursue peace in order for them to secure their lives. The solution to the problem of order in nature is to create a society in which men transfer their individual rights to a third party, the state, which creates the conditions of general stability. Society is thus based on a social contract by which members transfer and relinquish individual rights in the interests of peace. The result of this contractual arrangement based on mutual consent is 'a body politic' which 'may be defined to be a multitude of men, united as one person, by a common power, for their common peace, defence, and benefit' (vol. 4, p. 122). The body politic is thus the artificial body which provides the framework within which the real bodies of men can find security and peace.

There are, of course, many types of political bodies – monarchy, aristocracy and democracy – but Hobbes's main criterion of government is that it should govern in such a way as to maximize security (Sabine, 1963). All government involves sovereignty and security requires that sovereignty is absolute and indivisible. There can be no divisions within the body politic and therefore it is imperative that the church should be subordinate to the state. The other problem Hobbes had to tackle was the possibility of division within the family. In *De Corpore Politico*, Hobbes argued that man has a natural right to his own body and this raised the question of parental dominion over children. Hobbes noted that there might be an argument that the mother has a greater right over the child than the father, but such a situation might bring about a division of sovereignty within the household. Hobbes consequently came to the conclusion:

> It is necessary that but one of them govern and dispose of all that is common to them both; without which, as hath been often said before, society cannot last. And therefore the man, to whom for the most part the woman yieldeth the government, hath for the most part, also, sole right and dominion over the children. (vol. 4, p. 157)

The stability of the body politic rests on the patriarchal household in which the covenant between man and wife secures domestic peace. Hobbes went on to claim that by nature men are superior to women and therefore, in a system of primogeniture and monarchical government, male children would be preferred to female offspring:

Seeing every monarchy is supposed to desire to continue the government in his successors, as long as he may; and that generally men are endued with greater parts of
wisdom and courage, by which all monarchies are kept from dissolution, than
women are: it is to be presumed, where no express will is extant to the contrary, he
preferreth his male children before the female. Not but that women may govern,
and have in divers ages and places governed wisely, but are not so apt thereto in
general, as men. (vol. 4, p. 160)

Unlike many other seventeenth-century theorists of patriarchy, Hobbes
treated social institutions as artificial corporations or institutions rather than
natural arrangements. He did however produce a characteristic theory of
patriarchy in which the stability of society rests on the nature of sovereignty
within the household where husbands have indivisible power over the wife,
children and servants. The power of husbands was thus analogous to the
power of kings. For Hobbes, the continuity of society was grounded in the
continuity of bodies, property and power.

It has been argued that Hobbesian philosophy was thoroughly materialist
and that his conception of an exact science, as a model for the science of politics, was taken from geometry. The individual body was a point within
political space and the motion of the body was conceived in terms of appetite
and aversion. The multitude of bodies, especially in a state of nature, had few
distinguishing marks: 'each individual appeared as an atom, somewhat different in composition but having the same general appearance, hurtling
across a flat social plane; that is, a landscape without any visible contours of
social distinctions to bar his path or predetermine his line of motion' (Wolin,
1961, p. 282). The problem of order resulted from the fact that these bodies,
if unchecked, would periodically collide, rather like stars in the firmament.
The solution, as we have seen, was the creation of a sovereign power to regulate the motion of bodies. The notion that Hobbes did not consider the
effect of social distinctions on the motion is not entirely correct, since
Hobbes placed certain bodies (those of children, women and servants) under
the control of patriarchal powers. Female bodies were, so to speak, slower
and less weighty than male bodies, because the former were less endued
with 'wisdom and courage'. Hobbesian philosophy was nevertheless essentially individualistic in that it could not offer an account of the ways in
which societies are structured by social class, ethnicity, status groups or gender. For Hobbes, sexual differentiation was simply a differentiation of bodies
and their potentialities; he had little conception of the cultural specialization
of men and women into social roles. Almost every aspect of Hobbesian
materialism is now open to question. It is difficult to maintain that
Euclidean geometry provides the basic map of material reality (Harré, 1964;
Peters, 1956). In addition, many of the assumptions which are necessary for
the theory of the social contract, such as the state of nature argument, are
difficult to maintain.

It would appear that Hobbesian materialism has little to offer modern
sociology as a theory of social order. Hobbes's social contract theory appears
to be merely a point of departure for debates about the relationship between

consensus and coercion in social relations. However, modern discussions of values, hegemony, legal coercion and economic compulsion as the basis of social order appear to have neglected the problem which was central to Hobbes, namely the problem of the body in space and time. In this chapter, I want to suggest that it is possible to rewrite Hobbes in order to produce a theory of social order which starts out from the problem of regulating bodies. Such a theory can include an analysis of patriarchy and power without embracing *in toto* Hobbes's mechanistic view of the body as matter in motion. It is no longer possible to accept Hobbes's definition of the body, since the body is simultaneously physically given and culturally constituted. In this respect, it is interesting to consider Husserl's comment on the body in his study of the origins of geometry:

> All things necessarily had to have a bodily character – although not all things could be mere bodies, since the necessarily co-existing human beings are not thinkable as mere bodies and, like even the cultural objects which belong with them structurally, are not exhausted in corporeal being. (Husserl, 1978, p. 177)

Hobbes's physicalist account of the body is obviously not able to take into consideration the subjectivity of the body and the embodiment of consciousness in corporeal being. The other limitation is Hobbes's atomistic treatment of the body as an individuated entity in time–space motion.

## Neo-Hobbesian Problem of Order

Given these limitations on the original Hobbesian formulation of the problem of social order, it is possible, however, to formulate a neo-Hobbesian version of the body which will transcend these inherent limitations of his Euclidean framework. Following Foucault (1981) it is important to make a distinction between the regulation of populations and the discipline of the body. Following Featherstone (1982), it is equally important to make a distinction between the interior of the body as an environment and the exterior of the body as the medium by which an individual represents the self in public. At least initially, these dichotomies are proposed as a heuristic device for constructing a general theory of the body and for locating theories of the body. At an empirical level, these four dimensions cannot be nicely separated, but this fact does not expunge the analytic value of the model. The theory can be presented diagrammatically, as shown in Figure 1. The argument is that the Hobbesian problem of order as a geometry of bodies has four related dimensions which are the reproduction of populations through time and their regulation in space, the restraint of desire as an interior body problem and the representation of bodies in social space as an issue concerning the surface of the body. In Parsonian terminology, every social system has to solve these four sub-problems. Since the government of the body is in fact the government of sexuality, the problem of regulation is in practice the regulation of female sexuality by a system of patriarchal power.

The reproduction of populations and the restraint of the body involves at the institutional level a system of patriarchal households for controlling fertility and at the level of the individual an ideology of asceticism for delaying sexual gratification in the interests of gerontocratic controls. The control of populations in space is achieved, as Foucault (1979) suggests, by a general system of disciplines with the generic title of panopticism. In essentials, such a system of control presupposes a bureaucratic registration of populations and the elimination of vagabondism. Finally, societies also presuppose a certain stability in the methods of self-representation in social space. In pre-modern societies, the individual body was represented through the impersonal and external *persona*, the mask which unambiguously defined its carrier. In modern societies, the problem of representation is particularly acute, since, partly as a result of the commodification of the body, the symbolic systems of presentation have become highly flexible.

|  | Populations | Bodies |  |
| --- | --- | --- | --- |
| Time | Reproduction Malthus Onanism | Restraint Weber Hysteria | Internal |
| Space | Regulation Rousseau Phobia | Representation Goffman Anorexia | External |

Figure 1

These four dimensions of the body have been considered by a variety of social theorists, but no single theory has yet attempted to present a coherent account of the relationship between these features of corporeality. However, to illustrate these dimensions it is possible to select a small group of social theorists who were especially associated with a particular feature of the corporeality of social relationships. For example, Thomas Malthus has been correctly identified with the debate about the reproduction of populations and the problem of population control through either natural or moral restraints. My argument is that Malthusianism was a potent ideology of the patriarchal household in a society where population growth was regulated by delayed marriage. Max Weber has been selected as the classic theorist of asceticism and its bearing on the moral regulation of the internal body. By way of a theoretical aside, it is also suggested that Weber, not Foucault, is the pristine analyst of social disciplines and the rationalization of the body. Two contemporary social thinkers are selected in respect of regulation and representation, namely Richard Sennett (1974) and Erving Goffman (1969). The spatial regulation of populations and the presentation of self via 'face-work' (Goffman, 1972) are problems of urbanized civilization. The locus of these features of social corporeality is to be found in the contradictions of intimacy and anonymity.

To illustrate further the complex texture of the body in society and society in the body, I want also to argue that certain characteristic 'illnesses' are associated with these dimensions of the body and that these 'illnesses' are manifestations of the social location of female sexuality, or more precisely 'illnesses' which are associated with subordinate social roles. The purpose of this classification is to make more explicit the analysis of medical history and sexual deviance which has been developed by Foucault (1973; 1981) by arguing that the medical problems of surbordinates are products of the political and ideological regulation of sexuality. Late marriage was a structural requirement of European societies until the late eighteenth century (Andorka, 1978), a requirement which was enforced by gerontocratic and patriarchal control. The demand for late sexual gratification was ideologically enforced by certain medical theories which proclaimed the physical dangers of sexual 'self-abuse' in onanism and which expressed middle- and upper-class anxieties about the threat of masturbatory insanity. Just as capitalists were encouraged not to spend their wealth in luxurious consumption, so dependents were encouraged not to spend their sexual potentiality in unproductive onanism. Similarly, hysteria in young women was the consequence of sexual unemployment, but a necessary feature of delayed marriage in a society where marriage was an economic contract. If hysteria in the pre-modern period was an illness of scarcity (namely the inability to create new households), anorexia in the twentieth century is an illness of abundance. Anorexia is the product of contradictory social pressures on women of affluent families and an anxiety directed at the surface of the body in a system organized around narcissistic consumption. Only a social system based on mass consumption can afford the luxury of slimming. Finally, if hysteria and onanism are, as it were, diseases of time, that is delayed time, anorexia and phobias are diseases of space, that is the location of the embodied self in social space; they are diseases of presentation. The most obvious illustration of this relationship between space and illness is agoraphobia which is literally the fear of the market place. Masturbatory insanity, hysteria, anorexia and agoraphobia are aetiologically illnesses of dependency, while their traditional diagnosis and treatment reinforced and legitimated patriarchal surveillance.

## Reproduction

Every society has to produce its means of existence (food, shelter, clothing) and every society has to reproduce its members. These two requirements were regarded by Engels (n.d., p. 6) as 'the determining factor in history', but the problem of populations has been largely ignored by Marxists. This theoretical silence is partly explained by Marx's violent rejection of Malthus as the 'true priest' of the ruling class and of Malthusianism as an explanation of ' "over-population" by the external laws of Nature, rather than by the historical laws of capitalist production' (Marx, 1974, vol. 1, p. 495n). While Marx claimed that every mode of production has its specific laws of population, he did not demonstrate how these laws operated in different epochs.

The result is that Marxist demography is very underdeveloped in relation to other branches of Marxist social theory. Marxism does, however, require a theory of population, since the production of the means of subsistence is intimately related to the reproduction of populations – a relationship which is the nub of Malthusianism. It has been argued that Marx, in fact, took the demographic history of the nineteenth century as a basic assumption of his analysis of capitalism. For example, the immiseration of the working class and the creation of a reserve army as a result of the displacement of labour by machinery have as an implicit assumption the stability of the fertility rate (Petersen, 1979). Furthermore, it is difficult to give an adequate explanation of patriarchy without taking into account the requirements of human reproduction and the relationship between population growth and household structure.

Malthus's argument against Condorcet and Godwin was published in his *An Essay on the Principle of Population* in 1789. Malthusianism had an elegant simplicity: efforts to improve the living standards of the poorest section of the working class above the level of subsistence would be self-defeating, since they would result in an increase in population. The increase in population growth would, by threatening the means of subsistence, restore the existing condition of poverty among the working class. For Malthus, humankind (or more precisely mankind) is dominated by two universal 'urges' – to eat and to satisfy the sexual passions – which he described as fixed laws of nature. These two urges stand in a contradictory relationship, since reproductive capacity always outweighs the capacity to produce food. The necessity to restrain the sexual urge in the interests of survival often leads to 'preventive checks' on population which are immoral – prostitution, homosexuality and abortion. Malthus's moral philosophy was, therefore, based on a sharp dichotomy between reason and passion. The unrestrained satisfaction of passion has disastrous consequences; indeed, any 'implicit obedience to the impulses of our natural passions would lead us into the wildest and most fatal extravagances' (Malthus, 1914, vol. 2, p. 153). In what he called 'some of the southern countries', the indulgence of the sexual impulse leads to a situation in which 'passion sinks into mere animal desire' (p. 156). Since sexual passion is necessary for reproduction, the solution is to be found in 'regulation and direction', not 'diminution or extinction' (p. 157).

There are three checks on population expansion beyond the means of subsistence, which are 'moral restraint, vice and misery'. The population will be reduced by starvation, by unnatural sexual gratification or by the exercise of reason to encourage moral control over population expansion. Given these choices, Malthus thought that, from the point of view of reason, it was desirable to bring about certain moral preventive checks rather than allow 'positive checks' such as war and famine to reduce the rate of reproduction. Malthus's view on celibacy and delayed marriage as the principal methods of prevention were influenced by his visit to Norway between the publication of the first essay and the revised version of 1803. In Norway, where market relations had not penetrated the agrarian subsistence economy, farmers could not marry until they possessed a holding of their own; marriage was controlled by economic relations so that a man could not marry until he could support a family. Farmers without land were forced to become servants in existing household units. Malthus thought that delayed marriage would provide the most rational system of population restraint, but it would also inculcate positive moral virtues. The time of delayed sexual gratification would be spent in saving earnings and thus lead to 'habits of sobriety, industry and economy' (p. 161). Malthusianism sought, therefore, not to abolish sexual passions, but, through reason, to redirect and regulate these necessary urges towards late matrimony.

Malthusian demographic theory has been criticized on a variety of grounds. As we have seen, Marx's criticism was that Malthus had derived population laws from fixed laws of human nature instead of treating 'instincts' as products of social relationships. Another criticism of Malthus is that he failed to see how technological changes in agricultural production could increase the food supply without any great increase in the cultivation of the land mass; in addition, technical changes in contraceptive methods provided the means of birth control within marriage without recourse to abortion. Partly in defence of Malthus, Petersen (1979) has argued that Malthus's emphasis on late marriage as a system of population control was, at least descriptively, a statement of the traditional European marriage system. The practice of late marriage among the peasantry was breaking down in Malthus's time and it was changes in marriage patterns which largely explained the increase in population in European societies in the eighteenth and nineteenth centuries. There is some agreement that the European marriage pattern, which combined late marriage and permanent celibacy for a large section of the population, was the principal social means for restricting fertility (Glass and Eversley, 1965). It is obvious that there are many competing explanations of 'the demographic transition', but family structure and marriage patterns appear to have played a major part (Laslett, 1972). A man could not marry unless, to use Laslett's expression, there was a vacant slot in the social structure which the new couple could fill. The word 'husband' itself derives from two words signifying 'to dwell' and 'house'; a husband was a householder who could afford to support a family without being a burden on the immediate community. It was not until the decline of subsistence farming, the growth of factory production and the emergence of urban occupations that the traditional pattern of late marriage began to decline in the working class. The collapse of the conventional system was accompanied by the growth of romantic love, the disappearance of parental supervision of marriage partners and the development of the modern nuclear family isolated from the wider kin (Shorter, 1977).

There are a number of highly technical debates which surround both Malthusianism and neo-Malthusianism, and there is a massive and growing literature on the sociology of fertility (Freedman, 1975). Many of these issues are not however pertinent to this present discussion. Malthus is important for my argument because his demography is deeply embedded in, indeed presupposes, a particular moral viewpoint. His analysis implicitly assumes the existence of patriarchy and gerontocracy, since the delayed marriage pattern which he seeks to support and maintain could not operate effectively without a system of patriarchal households. In turn, this system of household power requires a powerful sexual morality advocating the benefits of delayed sexual gratification and this morality was grounded in Christian theology. Malthus provides two arguments against 'vice'. First, moral deviation in the form of homosexuality, abortion and masturbation is simply contrary to Christian teaching, but such an argument from tradition is not entirely persuasive, especially for anyone who simply does not accept traditional Christian values. Malthus had a second line of argument which could be described as ethical utilitarianism: we will be happier in marriage if we arrive at that condition with our passions intact and our sexual energies undiluted. Sexual asceticism before wedlock is a period of moral accumulation prior to consumption within marriage. It was for this reason that masturbation came to be seen as an unproductive activity and a wasteful luxury of the morally idle.

Masturbation became an object of severe moral condemnation in the second half of the eighteenth century (Shorter, 1977). In previous centuries, there was often a relaxed attitude on the part of parents towards juvenile masturbation; indeed, some medical treatises encourage moderate masturbation as a method of achieving a balance within the body's fluids. One indication of a change in attitudes was the anonymous publication of *Onania or the Heinous Sin of Self-pollution* in 1710, which became a widely read tract. The author argued that a variety of maladies, both physical and moral, resulted from this practice. In 1758, Dr Simon-Andre Tissot published his famous medical treatise on onanism, suggesting both that it resulted in dire physical consequences and that it was largely incurable (L. Stone, 1979). In France and Germany, similar tracts appeared as in, for example, S.G. Vogel's *Unterricht für Eltern* of 1786, in which infibulation of the foreskin was recommended as one cure for masturbation. By the nineteenth century, there emerged a cluster of medical categories – primarily 'masturbatory insanity' and 'spermatorrhoea' – to classify the negative consequences of 'unproductive' sexuality (Engelhardt, 1974). Masturbation was held to be responsible for 'headache, backache, acne, indigestion, blindness, deafness, epilepsy and, finally, death' (Skultans, 1979, p. 73).

There is no evidence of the 'real' incidence of masturbation in pre-modern societies; what we do possess is some impressionistic evidence about the level of anxiety expressed by parents, doctors and clergy about its undesirable consequences. What is the explanation for this moral panic? One argument suggests that the more male children from the middle class left home to attend boarding schools, the more parents felt their loss of control over the moral development of their offspring. In the new public schools of England, it was feared that children would come increasingly under the dubious moral influence of their peers and their school-masters (Ariès,

1962). This change in childhood training was also associated with a shift towards an urban life-style (Shorter, 1977), but it was also connected with a new emphasis, especially in Protestant societies, on the fundamental importance of character-training in children (Grylls, 1978). Foucault (1981) regards the increased interest in masturbation in the nineteenth century as part of a general medicalization of the urban population, which came increasingly under the surveillance of medical institutions and professionals. Perhaps the most promising explanation is provided by L. Stone (1979), who argues that in the middle class parental anxiety may

> have been encouraged by the rising median age of marriage, rising fears that masturbation was on the increase. More and more men were spending a longer and longer part of their sexual mature years with no other outlet for their libido but masturbation or prostitution. (L. Stone, 1979)

For a longer historical standpoint, masturbation had always been regarded, at least in official and orthodox circles, as a major sin in both Christianity and Judaism (G.R. Taylor, 1953). In England, the Protestant Reformation brought with it not only a greater emphasis on personal sin, but a new view of the importance of childhood training and the duties of fatherhood. In the eighteenth and nineteenth centuries, however, patriarchal control over the household was to some extent weakened by the doctrine of individualism, the growth of public schooling and the slow decline of arranged marriages, which were inconsistent with the Puritan notion of individual responsibility. The horror over masturbation was a defensive reaction against what was perceived as a diminution of parental authority. In addition, and contrary to L. Stone (1979, p. 321), there was a close symbolic parallel between wasted seed and wasted capital. 'Selfpollution' was a secret and deviant practice which was a product of the control over reproduction under a system of monogamy and late marriage. It was also, within the Malthusian scheme of population control, a denial of character-building asceticism, which was regarded as a necessary adjunct of successful capital accumulation.

## Restraint

The reproduction of population has been in traditional European societies controlled by a variety of institutional means and especially by monogamy, celibacy, delayed marriage and patriarchy. The weakness of Malthus's argument, apart from its dubious moral basis, was that it often failed to examine the relationship between social class and reproduction. While all societies have to reproduce themselves, Engels in *The Origin of the Family* saw more clearly than Malthus that the working class reproduces labour, and the ruling class, inheritors of capital. In a system of primogeniture, the ruling class demands, at the personal level, a number of ascetic restraints over the sexuality of the household members in the interests of capital accumulation and conservation. The sexuality and reproduction of labour, at least in early capitalism, was restrained by Malthusian checks, especially disease and poverty.

Capitalism is, however, a combination of contradictory forces, as Marx constantly asserted. Individual capitalists have a strong interest in the health, reliability and discipline of their own workers – hence the capitalist's tolerance, if not enthusiasm, for evangelical Protestantism (Pope, 1942; Thompson, 1963). Individual capitalists do not, however, want the burden of Poor Laws, asylums and welfare taxation – hence the capitalist's interest in a 'reserve army' of labour and migrant workers. The brutal simplicity of Malthus's argument is thus apparent: where workers fail to exercise 'moral restraint' over their reproductive potential, they will be driven by poverty and misery to restrain their reproduction. The significance and meaning of the relationship between asceticism and capitalism are thus different for different social classes.

Max Weber's account (1965) of the connection between religious asceticism and capitalism is notorious, and equally subject to unflagging criticism (Eisenstadt, 1968; Marshall, 1982). Weber's thesis has often been rejected out of hand by Marxist critics as a myth which suggests that thrift is the origin of accumulation (Hindess and Hirst, 1975). This myth had been wholly destroyed by Marx's argument that primitive accumulation had been achieved by violence, especially in the form of enclosures which forced the peasant off the land. Against this criticism, it can be argued that Weber's Protestant ethic thesis presupposes the separation of the worker from the means of production as a necessary requirement of capitalism (Turner, 1981). Weber then asks, assuming the alienation of the worker from productive means, what else contributed to capitalist growth by encouraging investment, limiting consumption and disciplining workers? The answer was that Protestantism through the idea of the 'calling' and ascetic disciplines brought about the origins of a process of rationalization that transformed European industrial culture. While Weber is often charged with a naive view of the connection between capitalist discipline and ascetic restraints, similar perspectives have also been put forward by Marxists. Marx in the Paris Manuscripts charged political economy with adopting self-renunciation as its basic thesis and argued that the theory of population rested ultimately on ascetic principles:

> If the worker is 'ethical' he will be sparing in procreation. (Mill suggests public acclaim for those who prove themselves continent in their sexual relations, and public rebuke for those who sin against such barrenness of marriage. . . . Is not this the ethics, the teaching of asceticism?) The production of people appears as public misery. (Marx, 1970, p. 152)

In the *Prison Notebooks*, Antonio Gramsci suggested that Protestantism in America, by achieving new standards of disciplined and regulated work, had paved the way for modern managerial techniques in Taylorism and Fordism. These managerial methods suppressed the 'animality' of man, training him for the regular disciplines of factory life. The interesting feature of Protestantism was that it involved self-discipline and subjective coercion rather than being an ideology enforced upon workers. Protestantism brought about a rational ordering of the body which was thus protected from the disruptions of desire in the interests of continuous factory production. Where the church failed to provide this puritanical discipline, the state filled the moral gap:

> The struggle against alcohol, the most dangerous agent of destruction of labouring power, becomes a function of the state. It is possible for other 'puritanical' struggles as well to become functions of the state if private initiative of the industrialists proves insufficient or if a moral crisis breaks out among the working masses. (Gramsci, 1971, pp. 303–4)

Gramsci treated the ascetic ordering of the body not only as a requirement of stable capitalist production, but as the moral origin of a process of industrial rationalization culminating in Taylorism and scientific management.

The real weakness of Weber's analysis of asceticism was that it failed to consider the distribution of ascetic practices by class and gender. This theoretical neglect is partly illustrated by the relationship between consumption and production. While Marx attempted to locate the crisis of capitalism in the production of commodities, the completion of the circuit of commodity-capital by consumption was also necessary for the realization of surplus-value. In the so-called under-consumptionist theory of capitalism, the crisis of the capitalist mode of production results from the fact that the demand for commodities is depressed by low wages. Against the under-consumptionists, it can be argued that consumption takes place when capitalists buy commodities such as machinery for productive purposes as part of their investment in constant capital (Mandel, 1962). There is individual consumption by workers, but this is merely to reproduce their labour-power through the purchase of clothing and food. Marx (1970, vol. 1, p. 537) took the view that 'All the capitalist cares for, is to reduce the labourer's individual consumption as far as possible to what is strictly necessary.' This argument against the importance of individual consumption appears, however, to be static and historically implausible, since it neglects the expansion in the productive capacity of capitalism through technical changes, improved management and the struggle of the working class to increase wages. Consumption in capitalism can either be confined to a narrow section of society (a 'consumption class') or be expanded through mass production to all classes (Hymer, 1972). This claim is not to deny that there is great inequality in consumption capacity or that the export of commodities plays a major part in the realization of surplus-value. The implication is that, in addition to ascetic denial of immediate consumption by capitalists in order to accumulate through further investment in productive capital, there must also be hedonistic consumption of goods if surplus-value is to be realized. It is this contradiction between hedonistic consumption and ascetic production which Weber failed to consider as a requirement of continuous capitalist development.

In the nineteenth century, consumption was restricted to a 'leisure class', but in the twentieth century a number of important changes took place which

facilitated the development of mass consumption. In the middle of the nineteenth century, the distributive system was underdeveloped and lagged behind
the system of industrial production (Jeffrey, 1954). The rise of consumerism
presupposes an urban environment, a mass public, advertising and the development of rationalized distribution in the form of the department store. In
Britain in the 1880s most of the conditions were eventually provided by the
transformation of retailing and distribution, along with the growth of advertising magazines. Other changes also had to take place in production such as
the standardization of commodities, which in turn made the advertising of
goods feasible in a context where commodities were replicated on a mass
scale. If the early department store played an important part in the development of a commercialized bourgeois culture, the supermarket has completed
this process of rationalization in the distributive system by making commodities available to a mass market of consumers (Miller, 1981;
Pasdermajian, 1954). Such a market context required asceticism at the place
of production in terms of Tayloristic management of the labour process, but
at the point of consumption it required a new life-style, embodied in the ethic
of calculating hedonism, and a new personality type, the narcissistic person.
Late capitalism thus involves a contradictory combination of asceticism and
hedonism, which are spatially differentiated between the factory and the
home.

Weber argued that there was an elective affinity between Protestant asceticism and the spirit of capitalism as exhibited in the works of Benjamin
Franklin. The notion that 'time is money' was the secular counterpart to the
Protestant concern that idle hands make easy work for the Devil. While this
relationship is plausible, there is also ample evidence that individual capitalists in their personal lives did not in fact conform to this ethic. Even Benjamin
Franklin appears to have diverged massively from this ascetic code (Kolko,
1961). Furthermore, when Weber referred to 'capitalists' he was of course
considering male capitalists. It is, therefore, important to examine the role of
social restraints of an ascetic nature on the body of women in the period of
early capitalism (Smith-Rosenberg, 1978). As in feudalism, early capitalism
required widespread restraints on female sexuality, especially among bourgeois women, to secure the stability of the system of property distribution.
The nature of these restraints is dramatically illustrated by the history of
female hysteria in the nineteenth century.

The Victorian notion of the 'hysterical woman' and earlier diagnostic labels
such as 'melancholy' and 'vapours' are to be explained in terms of the contradictory social pressures on women. The term 'hysteria' is derived from the
Greek word *hystera* or 'womb', since the cause of hysteria in classical medicine was thought to be the under-employment of the womb. In Egyptian
medicine, the womb was thought to dry out unless the woman was regularly
pregnant and, by floating upwards in the body, caused pressure to build up on
the brain. In Galenic medicine, the female seed becomes corrupt if it is not
fertilized and this putrefaction produces the hysterical outburst (Veith, 1965).
In the seventeenth century 'melancholy' was considered to have a similar

aetiology. For example, Robert Burton in *The Anatomy of Melancholy* of 1621 noted that working women rarely suffered from melancholy, while wealthy but unmarried women were commonly oppressed by the malady. His solution was marriage, religion and suitable occupations, such as charitable pursuits among the poor. What we might call the lazy womb as a physiological state was thus correlated with the lazy person as a moral condition, prevalent among certain classes of women. The social restraints of marriage were required to promote the mental stability and personal happiness of women. Women were, however, caught in a contradictory set of circumstances. They were regarded as overcharged with sexual energies, but marriage, as the only legitimate outlet for their sexuality, was often delayed within the European marriage pattern. In addition, those women who delayed marriage in the late Victorian period in order to follow a career in teaching or nursing were assumed to be especially exposed to the threat of hysterical breakdown. While parents worried about masturbationary insanity in boys, there was also anxiety about the dangers of female masturbation in a system of delayed marriage. Both masturbation and hysteria had a common root in the spoiled child: 'Petted and spoiled by her parents, waited upon hand and foot by servants, she had never been taught to exercise self-control or to curb her emotions and desires' (Smith-Rosenberg, 1972, p. 667). The answer to the sickness lay in self-discipline and good works under the watchful regime of parental restraint.

Once inside marriage, however, women were thought to be sexually underdeveloped, if not frigid, and it was this situation which drove men to prostitutes, while also excusing their behaviour. The transformation of the passions in women from adolescence to marriage was absolute, albeit somewhat miraculous. While during pregnancy they avoided the horrors of hysteria, women were confined to a private domestic sphere, where isolation and the burden of children brought on new forms of depression. The problem was that men were both necessary for female happiness and, through endless pregnancies, the cause of their distress. In the words of a more recent study of sexuality, we are reliably informed, by a man, that masturbation in women 'is always abnormal' and that 'the woman's sexuality remains dormant until it is awakened by a man' (Schwarz, 1949, p. 43). Of course, this also had to be the 'Right Man' rather than any man, since a woman had to accumulate her energies for lawful procreation. Thus, hysteria as part of a medical ideology of true womanliness had the social functions of keeping women in their place, that is the privacy of the domestic sphere away from the dangers of public life.


## Regulation

It is difficult to separate the problems of reproduction and restraint from the growth of an urban society in which populations were regulated in social space. From the eighteenth century onwards, urbanism was seen increasingly as a threat to culture, especially to the dominant culture of the elite. The growth of industrial cities involved the collapse of the traditional system of 'appearential ordering' whereby persons had been defined by the visibility of fixed status (Lofland, 1973). The techniques of regulation came, in social theory, to be bound up with questions of interpersonal intimacy and social anonymity, which in turn gave rise to a new input into the traditional Hobbesian social contract. The nature of population densities and their impact on character-structure became a linking theme in French social theory from Rousseau to Lévi-Strauss.

Unlike Hobbes, Rousseau's account of civil society was much exercised by the problems of urban existence. In Rousseau's general philosophy, human solitariness was taken to be a basic moral principle which provided the normative perspective for his treatments of nature, education and religion. The negative effect of urban crowding was to make men too dependent on the opinion of others, and their proper self-respect (*amour de soi*) degenerates into selfishness (*amour-propre*). In the discourse 'On the origin and foundation of the inequality of mankind', Rousseau sought to draw a clear contrast between the autonomous savage ('solitary, indolent and perpetually accompanied by danger') in a state of nature with urban man in civil society:

> *Amour-propre* is a purely relative and factitious feeling, which arises in the state of society, leads each individual to make more of himself than of any other, causes all the mutual damage men inflict one on another . . . in the true state of nature, *amour-propre* did not exist; for each man regarded himself as the only observer of his actions, the only being in the universe who took any interest in him, and the sole judge of his deserts . . . he could know neither hatred nor the desire for revenge, since these passions can spring only from a sense of injury. (Rousseau, 1973, p. 66n)

The problem of society is the problem of public comparisons and our dependence on social rather than personal reputation. Entry into society, especially into a city existence, obliterates pity which is mankind's only 'natural virtue'. The more people live in the company of others, the more selfish their behaviour becomes, since urbanization undermines natural compassion. In short, Rousseau argued that 'In proportion as the human race grew more numerous, men's cares increased' (p. 77). Troubles accumulate with the accumulation of men in urban space.

This inverse relationship between the quality of moral life and the quantity of urban bodies was also the basis of Rousseau's views on the theatre in the controversy with M. d'Alembert. In Rousseau's letter on the theatre, he was concerned to contrast the effects of theatrical performances in Geneva and Paris. In the urban environment of Paris, where the citizens are already corrupted by *amour-propre*, the theatre functions as part of state policy to entertain citizens who have nothing more positive to do with their civil liberties. By contrast, in the small republic of Geneva, the theatre must necessarily corrupt free men by exposing them to 'civilization'. In the large city 'everything is judged by appearance because there is no leisure to

examine anything' (Rousseau, 1960, p. 59). Because the citizens are contaminated by selfishness, reputational worth rather than personal value
becomes the sole criterion of personal stature. The theatre encourages reputational prestige, especially among women who adorn their bodies in a
competitive struggle for public attention. In the crowded spaces of urban
society, interpersonal familiarity breeds contempt. This theme was the dominant aspect of Rousseau's final publication, namely *Reveries of the Solitary
Walker* (1979). In the ninth walk, Rousseau observed that when strangers
first meet there is a formal courtesy expressed between them, but as these
strangers become more familiar, civility begins to disappear. Intimacy and
respect seem mutually exclusive. Public formalities appear to be necessary in
the densely populated spaces of the industrial city, but they break down
under the pressure of reputational displays and false selfishness. The innocence of free space disappears with the emergence of urban society; the
transition from

> nature to culture depended on demographic increase, but the latter did not produce
> a direct effect, as a natural cause. First it forced men to diversify their modes of
> livelihood, in order to exist in different environments, and also to multiply their
> relations with nature. (Lévi-Strauss, 1969, p. 173)

The density of populations produces an extension and intensification of the
social division of labour, which in turn binds people together in reciprocal
relations, thereby creating greater mutual dependency.

These themes in Rousseau's view of the state of nature were reproduced in
Durkheim's analysis of the division of labour (1964) and also in the romantic perspective of Lévi-Strauss's *Tristes Tropiques* (1976). Lévi-Strauss's
autobiographical commentary on anthropology can be read as a Rousseau-
like analysis of the consequences of Western, urban culture on primitive
simplicity. In his first encounter with the West Indies, he observed 'This
was not the first occasion on which I have encountered those outbreaks of
stupidity, hatred and credulousness which social groups secrete like pus
when they begin to be short of space' (Lévi-Strauss, 1976, p. 33). At a later
stage, he was forced to note the distinction between the solitude of the South
American forests and the human misery which characterized the densely
populated space of Indian cities. The cities of the Indian subcontinent
secreted 'Filth, chaos, promiscuity, congestion, ruins, huts, mud, dirt; dung,
urine, pus, humours, secretions and running sores' (p. 169). For Durkheim,
population density and the division of labour result in a society based on
reciprocity (organic solidarity) in which the individual is less subject to collective culture (*conscience collective*). For Rousseau and Lévi-Strauss,
urbanization and population density undermine the moral coherence and
dignity of the individual. In this respect they articulated a persistent *motif* of
nineteenth-century social thought, namely an anxiety about the moral
consequences of urbanization.

The Hobbesian solution to the problem of order in the theory of the social
contract started out from the premise of the materiality of single bodies; the

sociological problem here is that of the multiplicity of bodies in an urban environment in which interpersonal moral checks are thought to have collapsed. In Rousseau's terms, urban familiarity engenders moral contempt. My argument is that the 'solution' to this dilemma can be seen in terms of Foucault's 'anatomo-politics of the human body' and the 'bio-politics of the population' (Foucault, 1981, p. 139). Urban bodies were politically dangerous without the web of institutional regulation and the micro-disciplines of control. The surveillance and supervision of urban populations were achieved through regulation and classification, which made possible the centralized registration of bodies for policing under a system of panopticism (Foucault, 1979). In both Weber and Foucault, there is the notion that populations become progressively subordinated to rational disciplines under a process of bureaucratization and rationalization. The dangers of urban space nevertheless remained an ever-present reality for nineteenth-century liberalism: 'Appalled at the ethic of a crowded industrialized society, with its "trampling" and "elbowing", and dismayed at the ugliness of urbanized civilization, Mill sought comfort in solitude and communion with nature' (Wolin, 1961, p. 323). Demographic pressures, economic scarcity and political instability were forces which were concentrated in the narrow streets of the European cities.

These anxieties were in particular focused on middle-class women, who were seen to be especially exposed to the sexual dangers of urban space. Although Rousseau had strong views on individual freedom – 'Man is born free; and everywhere he is in chains' (Rousseau, 1973, p. 1165) – he assumed that women, as guardians of private morals, would be securely located in the domestic sphere (Okin, 1980). Women were especially susceptible to the dangers of false self-regard; as we have seen, theatres encouraged women to decorate their bodies in reputational competitions. In the city, new dangers abounded: infatuations, insults, abduction and moral degradation. The woman who stayed at home away from such dangers and temptations was both displaying the economic status of her husband and proclaiming her moral innocence:

> Women appearing in the streets alone 'had to be' women who went working of necessity, women whose husbands could not provide for their families single-handedly; such women could not possibly be decent. . . . Her domesticity demonstrated her economic and erotic dependence on her husband, and this in turn proved that he could provide for her material and erotic needs. (de Swaan, 1981, p. 363)

When the conditions which made the streets safe for women – street lighting, a police force, reduction in street violence – had been developed by the end of the nineteenth century, male anxiety about female independence necessarily increased. At this point, the first coherent medical description of agoraphobia appeared in 1872. The agoraphobic syndrome has not changed since the 1870s, being simply defined in terms of an anxiety about leaving the home, visiting shops, travelling alone or entering crowded spaces. In Freudian terms, the agoraphobic fears sexual seduction and represses libidinous interests in strangers. Agoraphobia in wives expresses the anxiety of husbands with regard to their control over the domestic household, but it also expresses the wife's dependence on the security and status of the bourgeois family setting. There is, therefore, a certain degree of collusion between partners as to the symbolic significance of the 'illness', which is reinforced by a professional interest in the reality of the complaint on the part of psychotherapists. The complaint both expressed female dependency and reproduced it. Fear of the market place had now been successfully converted into a medical condition, which legitimated the power relationships of the household.

Urbanization threatened the code of impersonal *civilité* with shallow intimacies, unregulated by respect for status and position. Paradoxically, the growth of intimacy entails a decline in sociability (Sennett, 1974). The courtly tradition of manners had permitted communal sociability between strangers by discouraging selfish expressions of intimate behaviour; intimacies are socially exclusive, but also express lack of genuine feeling (Weitman, 1970). By contrast, a secular urban society generates a cult of intimacy and affectivity between strangers which offsets the threat of anonymity and which attempts to deal with public space by replacing courtly values of impersonal *civilité* (Elias, 1978). In the nineteenth century, anxieties about seductive intimacies between anonymous strangers found their symbolic expression in female agoraphobia (Sontag, 1978). As women from the middle classes entered public society in the twentieth century with the growing demand for labour in the war-time crisis of Western capitalism, 'female complaints' became increasingly presentational and symbolic of anxieties about the surface of the body. For example, dietary practices were no longer aimed to control passions within a religio-medical framework; they are now aids to self-presentation in a context where ageing is no longer expected to preclude our capacity for presenting a good face.

## Representation

In pre-modern societies the person was housed in the *persona*, a public mask which was impersonal and objective (Mauss, 1979). Personality was objectified in the external marks of status and insignia. In feudal times, personhood and dignity came to reside in a man's shield, which was a privilege indicating class position. With the development of the surcoat, lambrequin and closed helmet, heraldric signs came to stand for distinction and were marks of identification of both person and status (Fox-Davies, 1909). In such a society, the moral value of a person was embraced by the notion of 'honour' which was embedded in institutional roles so that personal and social symbols coincided. This world of honour was transformed by the development of capitalism. In England, the aristocracy was largely demilitarized by the seventeenth century and, with the enclosure movement, was transformed into an agrarian capitalist class: 'The idiosyncrasies of the English landowning class in the epoch of Absolutism were thus to be historically interlocked: it was unusually civilian in background, commercial in occupation and commoner in rank' (P. Anderson, 1974, p. 127). The hierarchical concept of honour by inheritance was gradually replaced by the notion of the gentleman as the product of education. The 'honourable gent' was urban, commercial and non-military; his status was achieved, but the commercial background was clothed with the culture of a private education (Ossowska, 1971). With the development of capitalism, formal differences on the basis of status within an hierarchical system have been overtaken, at least in principle, by differences of merit and achievement so that personal worth can no longer be invested in external signs. In practice, status symbols denoting personal worth – in housing, speech, dress and other consumption patterns – persist, but these symbols are not exclusive rights with the backing of legal entitlement. Personal moral status has become more fluid, open and flexible; the modern personality now has dignity rather than honour: 'The concept of honour implies that identity is essentially, or at least importantly, linked to institutional roles. The modern concept of dignity, by contrast, implies that identity is essentially independent of institutional roles' (Berger, 1974, p. 84). The self is no longer located in heraldry, but has to be constantly constituted in face-to-face interactions, because consumerism and the mass market have liquidated, or at least blurred, the exterior marks of social and personal difference.

The extension of the franchise and the growth of mass consumer markets have facilitated the disappearance of ascriptive signs of personal value. Although hierarchical differences at work are crucially important for personal status, mass entertainment and the leisure market are relatively free of social exclusion based on class. The commercialization of sport has reduced traditional class differences both within and between particular sporting activities. In leisure styles, the universality of jeans and T-shirts does not remove class distinctions, but it does mask them behind the informality of dress. Variations between societies are clearly important. The English bowler hat is still symbolic of class and personality, whereas the Australian summer enforces a certain stylistic egalitarianism: 'there is a real sense in which the absence of clearly visible and unambiguous marks of inferior status has made the enforcement of an all-pervasive deference system almost impossible to sustain outside the immediate work situation' (Parkin, 1979, p. 69). Self and the presentation of self become dependent on style and fashion rather than on fixed symbols of class or hierarchical status. Urban space becomes a competitive arena for presentational conflicts based on commercialized fashions and lifestyles. There is a sense in which the self becomes a commodity with an appropriate package, because we no longer define ourselves exclusively in terms of blood or breeding.

This world of the performing self has been theoretically encapsulated in a number of streams of American sociology, particularly in so-called symbolic interactionism. Sociological awareness of the new personality structure of consumer society can be traced back to a number of classic texts. The concept of the social self in the American tradition of sociology is redolent of the naked space of consumer society. In *Human Nature and the Social Order*, Cooley (1964) spoke of the 'looking glass self' which cannot exist outside the gaze of others; our appearance in the mirror of others' responses was seen to be not only the basis of personal esteem, but constitutive of the self. Within social interactionism, the self and our public appearance are not so much conjoined but merged (G. Stone, 1962). The importance of the presentational self can be charted in Whyte's 'organization man' (1956), Fromm's 'market-oriented personality' (1941) and Riesman's 'other-directed personality' (1950). The tradition culminates in the contemporary debate on the 'narcissistic personality' (Lasch, 1979). The theme of these commentaries on American life is essentially Rousseauist: suburban America produces what Riesman called the 'lonely crowd' within which egoistic actions are draped in a false intimacy. My argument is that these texts are simultaneously diagnostic and symptomatic – they grasp the social disease of self-regarding intimacy while also expressing it. This feature of American sociology found its epitome in Goffman's compendium of interactionist concepts – 'face-work', 'deference and demeanor', 'stigma' and 'expression games' (Goffman, 1968; 1970; 1972).

Goffman's most influential work was *The Presentation of Self in Everyday Life* (1969). In Goffmanesque society, social relations constitute a stage, upon which the social actor presents a performance either individually or in the company of a team. These social performances are threatened by the possibility of perpetual failure; performances may be disrupted by forgotten lines, embarrassment, misinformation and discrepancy. The ritual order of everyday encounters is precarious and in need of constant repair. In terms of the Hobbesian problem of order, social actors are primarily motivated by self-regard and by the desire to maintain their 'face' at all costs; order exists insofar as social actors seek to avoid stigmatization and embarrassment in public gatherings. Social life is a game in which there is little scope for trust, since all human action is simply bluff and counter-bluff. Survival in this competitive world of social espionage hinges simply on the ability to select the most advantageous set of interpersonal tactics. Goffman's dramaturgical model is thus both a mode of understanding the new middle class and a reflection of its values:

> The dramaturgical model reflects the new world, in which a stratum of the middle class no longer believes that hard work is useful or that success depends upon diligent application. In this new world there is a keen sense of the irrationality of the relationship between individual achievement and the magnitude of reward, between actual contribution and social reputation. It is the world of the big-priced Hollywood star and of the market for stocks, whose prices bear little relation to their earnings. (Gouldner, 1971, p. 381)

Society as theatre is thus Rousseau's vision of urban *amour-propre* taken to its logical conclusion – a society in which reality becomes entirely representational.

Social success depends on an ability to manage the self by the adoption of appropriate interpersonal skills and success hinges crucially on the presentation of an acceptable image. Image-management and image-creation become decisive, not only for political careers, but in the organization of everyday life. In turn, successful images require successful bodies, which have been trained, disciplined and orchestrated to enhance our personal value. A new service sector of dietitians, cosmetologists and plastic surgeons has sprung up to augment the existing body-work professions of dentistry, hair-dressing and chiropody. In the managerial class, in order to be successful it is also important to look successful, because the body of the manager is symbolic of the corporation. The new ethic of managerial athleticism is thus the contemporary version of the Protestant ethic, but, fanned by the winds of consumerism, this ethic has become widespread throughout the class system as a life-style to be emulated. The commodified body has become the focus of a keep-fit industry, backed up by fibre diets, leisure centres, slimming manuals and outdoor sports. Capitalism has commodified hedonism and embraced eudemonism as a central value:

> The 'revolution in manners and morals', which took shape in the twenties when capitalism began to outgrow its dependence on the work ethic, has eroded family authority, undermined sexual repression and set up in their place a permissive hedonistic morality tolerant of self-expression and, the fulfilment of 'creative potential'. (Lasch, 1979, p. 45)

The new hedonism does, however, have peculiar features. It is not oppositional, being perfectly geared into the market requirements of advanced capitalism; it is heavily skewed towards the new middle class; it is also compatible with asceticism. Hedonistic fascination with the body exists to enhance competitive performance. We jog, slim and sleep not for their intrinsic enjoyment, but to improve our chances at sex, work and longevity. The new asceticism of competitive social relations exists to create desire – desire which is subordinated to the rationalization of the body as the final triumph of capitalist development. Obesity has become irrational.

All illness is social illness. At a trivial level, we know that stress is an important element in the aetiology of much chronic illness and that stress is the product of the temporal rhythms of modern societies; social stress results in peptic ulcers (Dossey, 1982). Illness also has social consequences in the form of unemployment and domestic disruption, but at a more fundamental level social processes constitute illness, which is a medical classification of a range of signs and symptoms (King, 1954). The meaning of illness reflects social anxieties about patterns of social behaviour which are deemed acceptable or otherwise from the point of view of dominant social groups. It has been argued that onanism and spermatorrhoea were medical categories which expressed the anxieties of parents whose authority over dependents was being questioned by new social arrangements. Hysteria was a metaphor of the social subordination of women, especially middle-class women who were attempting to express their individual independence through professional employment. Agoraphobia symbolized the uncertainty of urban space; fear of the market kept women at home, but also confirmed the husband's economic capacity to maintain a domesticated wife. If the argument is correct that in late capitalism there is for the individual a representational crisis of self-management, then we might expect, especially for women, the emergence of a presentational illness. In Goffman's dramaturgical metaphor, the characteristic illness for women should be bound up with the anxieties of face-work; it is anorexia nervosa which most dramatically expresses the ambiguities of female gender in contemporary Western societies. While it would be futile to deny that anorexia has psychological and physiological features, it also has a complex sociological aetiology and is deeply expressive of the modern view of beauty as thinness (Polhemus, 1978).

While I have attempted to separate certain illnesses in terms of reproduction, restraint, regulation and representation, the illnesses of women have one important thing in common – they are, at least sociologically, products of dependency. Female sickness – hysteria, depression, melancholy, agoraphobia, anorexia – is ultimately a psychosomatic expression of emotional and sexual anxieties which are built into the separation of the public world of authority and the private world of feeling (Heller, 1979). Masturbatory insanity and hysteria are not 'diseases' but deviant behaviour which express a crisis of delayed time: the problem of waiting for maturity in the transition from one household to another. Agoraphobia and anorexia are expressive of the anxiety of congested space. The agoraphobic suffers from protective patriarchy, the anorexic from protective parenting in the confines of the privatized family. As diagnostic categories, these illnesses also express male anxieties about the loss of control over dependants as women left the household for work and were allegedly exposed to public seductions.

The Hobbesian problem of order was historically based on a unitary concept of the body. The social contract was between men who, out of an interest in self-preservation, surrendered individual rights to the state, which existed to enforce social peace. However, the regime of political society also requires a regimen of bodies and in particular a government of bodies which are defined by their multiplicity and diversity. The Hobbesian problem is overtly an analysis of the proper relationship between desire and reason, or more precisely between sexuality and instrumental rationality. This problem in turn can be restated as the proper relationship between men as bearers of public reason and women as embodiments of private emotion. When expressed in this fashion, it is heuristically useful to identify four sub-issues within the general problem of order. The value of the model is that it brings into focus the fact that all social structures which institutionalize inequality and dependency are fought out at the level of a micro-politics of deviance and disease. Because the body is the most potent metaphor of society, it is not surprising that disease is the most salient metaphor of structural crisis. All disease is disorder – metaphorically, literally, socially and politically.

# 5

# Patriarchy: Eve's Body

Of Man's First Disobedience, and the Fruit
Of that Forbidden Tree, whose mortal taste
Brought Death into the World, and all our woe,
With Loss of Eden

Milton, *Paradise Lost*, Book 1

My argument is that any sociology of the body will hinge ultimately on the nature of the sexual and emotional division of labour. The sociology of the body turns out to be crucially a sociological study of the control of sexuality, specifically female sexuality by men exercising patriarchal power. There are two conventional explanations of the social subordination of women, which turn out, on closer inspection, to be in fact one argument. The first may be called the nature/culture argument and the second, the property argument. One feminist account of the universality of patriarchy as a system of power relations of men over women is that, because of their reproductive role in human societies, women are associated with nature rather than culture and hence have a pre-social or sub-social status. Women have not, as it were, made the transition from animality to culture, because they are still tied to nature through their sexuality and fertility. The universality of women's subordinate status in society is thus explained by the universality of women's reproductive functions. The subordination of women is not essentially a consequence of physiology, but of the cultural interpretation of female reproductivity as denoting an unbreakable link with nature. The distinction between 'nature' and 'culture' is, of course, itself a cultural product. It is a classificatory scheme which allocates women to an inferior 'natural' category and men to a superior 'social' category.

## Nature/Culture Argument

The result of this association with nature is that men are seen to be liberated from natural functions in order to occupy themselves with higher status activities, namely the creation of a cultural, symbolic environment (Ortner, 1974). In this division of labour, men create enduring symbols, while women reproduce perishable bodies. The social roles of women come as a consequence to be seen as inferior to the social roles performed by men. The final step in the argument is that women are allocated and trained into a psychic structure ('maternal instincts', 'affection' and 'emotions') which is sharply opposed to the psychic space ('reason', 'reasonableness' and 'reliability') of men. This dichotomy between reason and desire is then associated with a further dichotomy between public and private space, such that women occupy the domestic world of private emotions and affections.

The nature/culture argument is clearly powerful and does offer a plausible explanation of the universality of patriarchal domination of men over women. The argument, however, suffers from a number of difficulties. In pre-modern societies, patriarchy typically involves the exercise of power by adult males over women, children and other dependent men. In Old Testament patriarchy, the tribal patriarch dominated both his wives and sons. To take a more contemporary illustration, Crapanzano (1973) provides an interesting analysis of the ambiguities in the social status of young men in a North African context. While men dominate women, it is also the case that fathers dominate their sons. Since masculine identity is conceived in terms of sexual strength, power over women and political dominance, unmarried males are forced into dependent social roles which are in many respects parallel to female roles. The consequence of this social location is that young men have a quasi-feminine personality. The psychosomatic illnesses which these young men experience are symbolic of their uncertain social status and are interpreted as consequences of spirit-possession by female demons (*jinn*). In empirical terms, it is difficult to separate out patriarchy from gerontocracy, namely the political dominance of male elders over household members of all generations and sexes. The early history of colonial America would be a further illustration of the employment of an ideology of biblical patriarchalism to buttress a system of gerontocratic government (Fischer, 1977). One solution for the nature/culture argument would be to claim that dependent men are, alongside women, identified with nature, but such a solution would weaken the original neatness of the thesis. In addition, it would be difficult to see in what sense young men are close to or part of nature.

A second weakness of the argument is precisely its generality (McDonough and Harrison, 1978). It is difficult to believe that societies have timelessly and universally interpreted women in the same nature/culture dichotomy. One alternative to this dichotomy was, for example, the notion that women were unnatural or monstrous creations and thus located somewhere between nature and culture. Aristotle regarded women as deviations in nature, but this Aristotelian view was often taken further to suggest that women were monsters outside nature (MacLean, 1980). Furthermore, it is not always the case that women are regarded as especially important in reproduction. I shall amplify this point later, but it is interesting to note that in one of the major mediaeval texts on reproduction – Giles of Rome's *De formatione corporis humani in utero* – it is specifically denied that the woman has any active part in the production of men. It is the male seed which is alone generative (Hewson, 1975). If women were not thought to be important in reproduction, how could they be thought to be close to nature?

The third weakness of the nature/culture argument is that while it may explain the origins of patriarchal attitudes, it is difficult to see how it could explain the maintenance and continuity of these attitudes. The nature/culture dichotomy becomes increasingly blurred and remote in societies characterized by urbanization, secularization and scientific medicine. 'Nature' is continuously and effectively appropriated by modern cultures. Advances in genetic technology – embryo transplants, sperm banks, artificial insemination, sterilization, contraception and prophylactic hysterectomy – mean that culture massively intervenes in natural processes. Indeed, what would count as 'natural reproduction' becomes increasingly uncertain and ultimately reproduction may pass entirely out of the realm of nature into the realm of culture. Women become increasingly separated from 'nature' by the intervention of 'culture' in the reproductive process.

## The Property Argument

The property argument suggests that patriarchal attitudes are an ideological outcrop of a more basic economic requirement, that is the regular distribution of property through legitimate heirs. Behind patriarchy, there lies the problem of paternity, namely the flow of property between generations according to male inheritance. This control of wealth through kinship requires both the control of wives and the control of children. The property argument is thus more pertinent to the explanation of patriarchy as simultaneously the control of wives and the control of dependent males. In pre-modern societies, women are seen as a potential threat to the solidarity of the kinship group, because there can be no absolute guarantee that the children they bear actually belong to the group. Given the presence of an incest taboo, men cannot take wives from their immediate kin. Because women are exchanged between families, the actual paternity of children is always open to a margin of doubt. As I shall show shortly, the practice of slaughtering all first-born males is at least one solution to this issue of legitimate offspring. This argument does assume, admittedly, the salience of biological over social fatherhood, but the emphasis on 'true' biological fatherhood does appear to be a social answer to the requirement of property distribution without competition from a variety of claimants who may not be genuine members of the kin. In a situation of economic scarcity, the cohesion of the household and its control over property are enhanced by the absence of sexual rivalries and disputes over paternity and legitimacy:

> Nowhere do we find unregulated, amorphous sexual promiscuity within the house, even if sexual relations between siblings are a recognized institution; at least nowhere on a normative basis. . . . Subsequent normative elaboration was obviously in the interest of safeguarding solidarity and domestic peace in the face of jealousies. . . . As a rule, then, a man acquires exclusive sexual rights over a woman when he takes her into his house or enters her house if his means are insufficient. (Weber, 1978, vol. 1, p. 364)

Patriarchal attitudes to women and sexual control over them are political and ideological arrangements which are based upon property distribution through particular forms of kinship relations. However, patriarchy as a power relationship also extends over younger men (especially first-born sons) whose sexuality must be controlled in the interests of household solidarity and economic stability. On this basis, it is clear that these two arguments about the universality of patriarchy could be combined, because the association of women with nature (and therefore with inferior status) then becomes a basis for legitimation of patriarchal control of property.

## Patriarchal Religions

Ideologies relating to the 'natural' character of women or to their need for protection can be treated as social resources which are mobilized by social groups, especially gerontocratic elites, to subordinate women, children and young men in the interests of property relations. These property relations within the traditional household form the economic basis of sexual ideology and interpersonal power. Interestingly enough, it is Christianity which, in the West, has formed the basis of patriarchal ideologies in both feudalism and early capitalism (Abercrombie, Hill and Turner, 1980; Turner, 1983). This connection between Christian theology and patriarchal power has been frequently noted. For example,

> Christian ideology has contributed no little to the oppression of woman. . . . [women] could take only a secondary place as participants in worship, the 'deaconesses' were authorized to carry out only such lay tasks as caring for the sick and aiding the poor. And if marriage was held to be an institution demanding mutual fidelity, it seemed obvious that the wife should be totally subordinate to her husband. (de Beauvoir, 1972, p. 128)

'Patriarchal religion' has thus been regarded as a major basis for the socialization of women into subservient and submissive roles (Millett, 1977; Seltman, 1956). While there is much justification for this interpretation, even in contemporary societies (Mercer, 1975), it is important to recognize that paternalism and patriarchy are systems of belief and practice which are located in the dynamics of particular household structures and that these households are determined by particular economic requirements. Unless this position is adopted, patriarchy appears to be a universal, free-floating essence which cannot be, as it were, tied down sociologically. It is equally important to recognize that the Christian doctrine was the product of several rather different traditions which were not fully crystallized into a coherent view of patriarchy until the institutionalization of casuistical theology in the thirteenth and fourteenth centuries.

The Christian view of women grew out of three sources: ancient Judaism, the Essene sect and Greek culture. Jewish society in Old Testament times was a confederation of tribes, bound together by a covenant with the one God Yahwe. Jewish tribes were collections of families claiming real or fictive descent from an ancestor or patriarch. These families were often polygamous. In Judges and Deuteronomy, it was recognized as a legal fact that men would acquire many wives and concubines within the family. In later periods, the Talmud attempted to fix the number of wives appropriate for different classes; kings were allowed up to eighteen, while a subject had only four (Vaux, 1961). The principal role of the wife was to be fruitful and multiply, and thus part of the pressure to acquire additional wives was associated with barrenness or low fertility. Similarly, a wife who gave birth to only female children would also be regarded as unsatisfactory. An unmarried woman was under the authority of her father, just as the married woman was under the authority of the husband. The husband was the master (*ba'al*) of the wife in the same way that he was *ba'al* over the fields. It is clear, therefore, that women as productive bodies were possessions of the head of the household alongside other possessions: servants, ox, ass and dwelling place. Although women had low status within the group, they were crucial to the reproduction of the family and the tribe. Further, since the tribe had a sacred status within a legal contract with God, the 'purity' of reproduction within the group came to depend on the purity of the women. With the break-up of the confederacy and the diaspora of the Jewish people, this requirement for pure Jewish reproduction was actually enhanced. Ultimately, to be Jewish was to have a Jewish mother (Yuval-Davis, 1980).

The moral coherence of the tribe was the guarantee of its social solidarity and these conditions were rooted in the sexual purity and fidelity of women. It is not surprising, therefore, that the regulation of women was strict to the point of brutality. Adultery between a man and a married woman was a property crime, that is, an infringement of the property rights of the injured husband. Such crimes were punished by death through stoning or burning. However, while men and women were encouraged to be faithful, a man taking a prostitute was regarded as dissipating his strength and wealth rather than his moral value since prostitution was not an invasion of property rights. This system of religious norms governing women's bodies expressed three group interests: the perpetuation of the family line, the conservation of domestic property and the preservation of the ancestral inheritance. These three interests necessarily dominated Jewish attitudes to children. Sterility in women was regarded as a divine punishment and this notion was combined with the belief that numerous sons were the symbol of patriarchal power. Among the male offspring, the eldest child was the most important, since he would become the head of the household and inherit the major share of the family property. At various points in the Old Testament, the principle of primogeniture was over-ruled by favouritism – Abel and Cain, Jacob and Esau. In the biblical story of Joseph and the coat of many colours, we see the principle of ultimogeniture (inheritance by the last-born male child) in operation in which Joseph was the son of Jacob's old age. These instances were, however, exceptions to the general rule of preference for first-born males.

This system of primogeniture under patriarchal domination did require confidence in legitimate paternity which was often difficult to guarantee. In a society where the precise facts of conception, gestation and reproduction were not clearly understood, there was always 'reasonable' doubt as to the paternity of first-born children. The patriarch could never be entirely certain that wives entering the household as young brides from other families were not already pregnant. In such a case, 'his' children would in fact be somebody else's. It would thus be possible for another family head to claim rights to property as a biological father. The themes of paternity and infanticide were consequently important to much of the mythology of the Old Testament world. The story of the patriarch Abram in many respects provides a summary of these themes. Abram's wife Sarai failed to provide children because of her barrenness and instead offered her Egyptian handmaiden Hagar as a concubine through whom Abram fathered a child called Ishmael. God then formed a covenant with Abram and this contract was symbolized by the circumcision of Abram and all the men of the tribe. Abram and Sarai changed their names to Abraham and Sarah. Furthermore, the covenant was validated by the fact that Sarah in extreme old age gave birth to a son called Isaac. As a test of the covenant God instructed Abraham to sacrifice Isaac by burning, but an angel intervened once Abraham had proved his willingness to slay his own son. With the death of Sarah, Abraham took another wife who provided him with many children.

The conventional interpretation of the circumcision ritual was that it was originally an initiation into marriage, but in Judaism it was a sign of incorporation into the group and hence into an alliance with Yahwe (Vaux, 1961). Since 'Abraham' meant 'the father of a multitude', the circumcision rite also stood for the willingness of other tribes to join in a covenant with God and the people of Israel (Epstein, 1959). The rite became a sign of inclusion and of exclusion (Douglas, 1970). There was, however, an alternative interpretation which suggested that circumcision was a symbolic alternative to infanticide. Because there was a suspicion that first-born male children may have been the product of sexual encounters prior to marriage between the wife and another father, one solution to this anxiety was to slaughter all such offspring. It has been suggested that the 'essence of Judaism and Christianity is the management of the infanticidal impulse' (Bakan, 1974, p. 208). Circumcision was symbolic of the slaughter of children by the knife; baptism, of drowning in water. In the circumcision rite, the 'children' of Yahwe as Father of the people were accepted into the community and the infanticide impulse was redirected, just as Abraham's intention to sacrifice Isaac was deflected to the ram. In psychoanalytic terms, circumcision was the reverse side of the Oedipus complex and the incest taboo. While Freud's view of religious practice in *Totem and Taboo* (1960) has been justifiably subject to criticism (Breger, 1981; Wollheim, 1971), Freud's clinical data importantly indicated the ambiguity of emotions and relationships between father and son, just as the circumcision rite pointed covertly to the ambiguity in the attitude of patriarchs to first-born males.

My argument is that one source of Christian ideology of women is to be located in Judaic social organization and in Jewish ritual. In Christianity, the inclusionary rite of circumcision was transferred to that of baptism, but there was also a transference of sacrificial and patriarchal symbolism. The vengeful God of the Old Testament was gradually transformed into the merciful Father of Christianity and the sins of men were expunged by Christ the

Sacrificial Lamb. Without accepting Freud's apparent commitment of the idea of Original Sin as an actual historic event, Freud's interpretation of Christian mythology is interesting, if not wholly convincing:

> In the Christian doctrine, therefore, men were acknowledging in the most undis-guised manner the guilty primaeval deed, since they found the fullest atonement for it in the sacrifice of this one son. . . . A son-religion displaced the father-religion. As a sign of this substitution the ancient totem meal was revived in the form of com-munion, in which the company of brothers consumed the flesh and blood of the son – no longer the father – obtained sanctity thereby and identified themselves with him. (Freud, 1960, p. 154)

As Breger (1981) points out, this interpretation by Freud presents a world in which women are strangely absent: it is a society of fathers, sons and broth-ers in which no space is allowed for mothers, daughters and sisters. In order to understand the place of women in Christian mythology and cosmology, we will have to turn from this ancient Judaic world to examine the impact of the Essene sect on the Christian treatment of the problem of reproduction and sexuality.

The Essenes were an ascetic Jewish sect which existed during the time of Christ and which came to an end with the suppression of the Jewish revolt in AD 70. Contemporary understanding of the importance of this sect has been greatly assisted by the discovery of the Dead Sea Scrolls at Qumran between 1947 and 1956 (Allegro, 1964; 1968; Dupont-Sommer, 1961). These Scrolls have given rise to various acrimonious debates about the relationship between the Essene sect and early Christianity. Most of these issues are not germane to my argument. The gist of the matter is that the Qumran sect, in its teach-ings and practices, anticipated much of the eschatological doctrinal core of early Christianity and in particular established an essentially negative attitude towards women as obstacles to the religious life. Essenism involved the shar-ing of communal property and the rejection of individual possession as unholy; it espoused a commitment to the fellowship of the human race, while expressing bitter hatred against its immediate enemies. The Essenes practised baptism and, while waiting in communal purity, prayed for the arrival of the teacher of Righteousness struggling against the Wicked Priest and the Man of Untruth to lead the community out of danger and restore the kingdom of God. During the period of preparation for the coming Messiah, the purity of the community depended on a strict regulation of sexual practices and the Scrolls spoke harshly about the temptations of the flesh and the seductive power of women. The Scrolls suggested that the lowly and dangerous status of women was connected with their reproductive role and hence with their closeness to nature. In this respect, the Essenes merely followed the traditional Jewish idea that menstruation created ritual impurity. However, they took this attitude further in regarding celibacy as a religiously prescribed institution. The adoption by Jesus of celibacy, at least after his reception of the Holy Spirit, may indicate the influence of Essenism on Christian practice. An alter-native interpretation is that in the Jewish religious tradition generally, it was held that prophecy was incompatible with marriage since the prophet must hold himself in readiness to receive the message of God (Vermes, 1976). The impact of both Essenism and Judaism on early Christian teaching was to define women as dangerous because of their natural impurity and to treat women as requiring close patriarchal supervision in the interest of legitimate property inheritance.

This negative view of women in early Christianity was amplified and mediated by the Greek context of Pauline theology and by the Greek reception of the New Testament. Paul's attitude to women has been frequently commented on in feminist literature (Figes, 1978; Mercer, 1975). Paul's discussion of sexuality and marriage in the First Epistle to the Corinthians has become notorious as an illustration of sexism:

> It is good for a man not to touch a woman. Nevertheless, to avoid fornication, let every man have his own wife, and let every woman have her own husband. . . . I say therefore to the unmarried and widows, it is good for them if they abide even as I. But if they cannot contain, let them marry: for it is better to marry than to burn. (1 Corinthians 7)

Similarly, when Paul was commenting on spiritual gifts, especially the gift of tongues, he recommended that 'women keep silence in the churches: for it is not permitted unto them to speak; but they are commanded to be under obedience, as also saith the law' (1 Corinthians 15). One justification for his negative view of women was derived from the Adamic myth in which the origin 'of all our woe' was located in the disobedience of Eve: 'For Adam was first formed, then Eve. And Adam was not deceived, but the woman being deceived was in the transgression' (1 Timothy 2). In selecting celibacy as a necessary basis for his calling as an apostle, Paul followed a traditional position by which religious vocations were incompatible with marriage, but Paul extended this belief over the whole community. The primary justification for marriage became a defence against fornication, because it was better to marry than to burn. Marriage was not so much a positive activity in itself but more the last line of defence against natural desire.

The Judaic view of women which was present in Paul's theology was reinforced in early Christianity by Aristotelian philosophy. Judaic Christianity, like the Essene sect, was primarily concerned with eschatology, not with rhetoric, but once Christianity developed in Greece it was forced increasingly to express belief within the garment of Greek philosophy. Eventually the Christian church itself became a hellenizing force by spreading to the urban commercial centres of the Mediterranean world. Christianity adopted the logic and the social organization of the classical world, and hence took over its attitude to women. In Aristotle's philosophy, the moral value of a person was defined by his function in society and, since he assumed that women were basically domestic workers alongside domestic slaves, their moral value was far below that of men (Okin, 1980). Because the privacy of the domestic sphere was regarded as deprivation by contrast to the freedom and rationality of the public sphere of politics, women were associated with necessity and toil. To be private was to be deprived and as a result women were not entirely suitable as companions for men. To understand this social arrangement, it is important to recognize that the idea of contest was central to classical life. Men were involved in two competitive spheres of the body (sport) and the mind (politics): women had no place in either. As a consequence, men sought fraternal contacts through homosexuality or through the system of courtesans (*hetairai*) (Gouldner, 1967). The validation of the male self thus took place outside the domestic sphere, leaving women locked within social roles of service and reproduction. While Paul condemned women to silence because of their association with seduction, the Greeks removed women from the world of discursive combat because their domestic character robbed them of public rationality.

Although under Roman law the position of women in society was somewhat alleviated (de Beauvoir, 1972), hellenized Christianity provided a powerful doctrine to legitimize the subordination of women both in society and in the church. This negative view of women was, I have suggested, closely connected with the nature of property distribution within the household, which was in turn associated with problems of paternity. The attempt to explain women's social exclusion in terms of their association with nature is secondary to these economic and political determinations. At the level of theology, however, the 'problem' of women was formulated in terms of the search for a rational solution to salvation. These ancient anxieties about women which grew out of the myth of Eve's seduction were increasingly formulated into an intellectualized theology and increasingly institutionalized in monasticism, celibacy and a priestly monopoly. Different religions can be conceived as distinctive salvational practices, but any rigorous attempt to achieve salvation must in Weber's view involve some solution to the irrational, excluded or sublimated, otherwise it conflicts with both the mystical and the ascetic path towards personal salvation. For Weber (1966, p. 239) Christianity 'went beyond all other religions in the limitations imposed upon permissible and legitimate sexuality'.

It is sometimes suggested that this problem of sexuality in Christianity split the female personality into diametrically opposed halves – either the harlot or the pure mother. Equally Christianity fractured love into *agape* or *eros*:

> In the Christian religion one finds what amounts to a total separation of spiritual and physical love, a renunciation of sexuality which is almost homosexual in its sentimental evocation of the pure mother figure and its emphasis on the union and companionship of a select band of brothers, the twelve Apostles. (Figes, 1978, p. 55)

It would be more correct to say that in Christendom there emerged a division of reproductive activity which divided the society into an elite which renounced sexuality and a mass which bore the burden of reproduction. This division cut across gender since both monks and nuns became virtuosi of spiritual love. There developed an exchange relationship between this spiritual elite and the mass which shouldered the necessary evil of reproduction. The elite performed a spiritual labour on behalf of those enmeshed in the duties of the flesh. These vicarious duties of the spiritual elite compensated for the dangers of reproduction which, while necessary, also gratified desire.

## The Feudal System

While this division provided a 'solution' for the necessity of reproduction, the orthodox system of Christian moral teaching never entirely matched the demographic and social requirements of household reproduction among the dominant class. This lack of perfect fit is well illustrated by the requirements of feudalism. In the feudal system, the accumulation of property in land pre-supposed a system of successful marriage alliances and the continuity of property between father and son. In turn, this system was aided by the pres-ence of a powerful doctrine of sexual behaviour in mediaeval Catholic doctrine. The stability of inheritance under either primogeniture or unigeni-ture was backed up by religious teaching which demanded female chastity, virginity in daughters, filial piety and duty among disprivileged siblings. The feudal system, while demanding pure wives, also required maximum fertility to guarantee male succession. In many respects, Catholicism answered both requirements ideologically by elevating the notion of the pure mother, while also encouraging the biblical imperative to 'go forth and multiply'. The reg-ular confession of women and the requirement of priestly control over confession which developed rapidly in the thirteenth century provided a pow-erful religious apparatus for the control of aristocratic women by lords through the intermediary of the priest (Foucault, 1981). The primary aim of confession was to protect orthodox belief and practice from heresy; the reform of the confessional in the thirteenth century was part of a wider reor-ganization of the church which was endangered by a variety of heresies and controversies. Confession sought to bring about public coherence and stabil-ity through the internal action of the individual 'court of conscience' (Hepworth and Turner, 1982). Although these confessional reforms may not have had the desired effect on the population as a whole, women within the dominant class were especially 'exposed' to the spiritual direction of the father confessor. The whole ritualistic and liturgical structure of the church gave formal backing to the moral theology of the Catholic church which, at the level of the feudal household, encouraged female submissiveness com-bined with maximum fertility. In principle, Catholic sexual teaching in the Middle Ages could thus be thought to be perfectly compatible with the economic requirements of feudalism.

There were, however, two permanent problems in the feudal system of monogamous unions and inheritance by primogeniture which conflicted with orthodox religious teaching. As we have seen, under the polygamous system of Old Testament patriarchy barren wives presented no major difficulty since they could be supplemented by additional women. Monogamy in feudalism presented a threat to inheritance in the case of barren wives, wives who pro-duced daughters or in situations where the male offspring died before their fathers. The mediaeval marriage system tended as a result to produce two contrasted patterns of family organization – the ecclesiastical and lay systems (Duby, 1978). The importance of the lay system was that it permitted the repudiation of barren wives; these repudiations were often legitimated by questionable appeals to incestuous unions. There was, as a result, a conflict between the economic importance of fertile wives and the religious requirement of life-long monogamy and fidelity. Thus the religious ideology gave only partial support to the economic interests of a patriarchal system.

The other tension in the feudal system was the social dislocation of unattached 'youths', namely the surplus population of younger sons of powerful feudal families. These younger sons found it difficult to form marriage contracts with women of their own class, because they could not inherit adequate entitlement to land. They formed instead casual alliances with household concubines or with peasant women or held adulterous aspirations towards married women. Their behaviour, unlike the ideal behaviour of the first-born male, was regarded as 'unruly' and hence they retained the title of 'youth' irrespective of their actual age and status. What defines 'youth' is not age, but the absence of binding commitments resulting from their precarious location in the social system (Davis, 1971; Smith, 1973). These 'youths' found an outlet for their passions in tournaments, war, chivalry and escapades. There was no space for such men within the church's view of the world which, for example, did not endorse tournaments or unregulated military adventures. One interesting feature of the unrestrained sexuality of these 'youths' is that it emphasized the fact that, at its basis, the marriage relationship was essentially an economic contract and hence marriage was somewhat remote from sexual love, affection and desire. It is for this reason that the 'youth' of mediaeval society have often been regarded as the main social carriers of courtly love which was in part an alternative to the dominant religious model of asexual love.

Courtly love as an ideal emerged towards the end of the eleventh century in Languedoc and developed as a dominant model of much mediaeval literature, achieving its apogee in *Le Roman de la Rose*. This tradition of sexual passion is associated with a class of landless, unattached knights for whom the lady of the castle represented an ideal, if distant object of their love. The main themes of courtly love poetry were humility, courtesy, adultery and the religion of love. As Lewis (1936) points out, these themes indicate a certain feudalization of love since they reflect the hierarchical court structure of feudal society. Romantic passion was channelled through the hierarchical status organization and it was directed upwards from the landless knight to the ladies of the court. Such love was expressed in terms of chivalry and courtesy, embracing a new etiquette of interpersonal manners. These romantic attachments tended to be adulterous precisely because these knights were excluded from appropriate marriages. The themes of courtly love poetry tended to be anti-Christian and there evolved a religion of love which was a parody of the central Christian virtues of chastity and virginity. This poetic tradition at least by implication recognized that marriage had nothing to do with love and that romantic attachments, to exist at all, could only be located outside wedlock. This interpretation of the poetic tradition of romantic love has been challenged by Robertson (1980) because, in the case of Chaucer, romantic love was seen to be destructive of the social fabric in which the stability of the family was the ultimate foundation of the stability of society. In some respects, however, Robertson's correction of the thesis of Lewis only serves to reinforce the point that courtly love was an oppositional tradition which recognized that the patriarchal organization of the household was an economic arrangement that precluded companionate marriage and attachment based on sentiment.

It has been argued that the household structure of mediaeval society required an ideology which enforced standards of fidelity and chastity in women as the basis for a system of inheritance based on male supremacy. This patriarchal system was a form of household power which subordinated both men and women to gerontocracy in the interests of property conservation. To some extent, Catholic moral theology was perfectly suited to these requirements, but the economic pattern of the household also gave rise to domestic concubinage, a stratum of riotous youth and an oppositional tradition of love poetry. It was partly out of this oppositional romanticism that a new appreciation of the individuality and subjectivity of women began to emerge, which was incompatible with much of the traditional conceptualization of women in society.

## Individualism

The social role of 'youth' in pre-modern society is important because it reinforces the argument that patriarchy is a system of domestic power which is exercised over both men and women of all age groups. The 'youth' is not a whole person, equipped with responsibilities and social power, because that person is removed from control over property. It could not be argued that 'youth' as a social category was somehow close to nature and therefore not part of the cultural core of society. These men were excluded from social status and power by virtue of their exclusion from property. On the other hand, this group of propertyless males had certain freedoms which were, at least normatively, proscribed to male first-born heirs. The eldest son in principle had to conserve his body so that his seed could be the carrier of household property and political power. In addition, the cultivation of courtly love poetry, especially in the troubadour tradition, resulted in a new conception of woman as a person capable of individual sentiment, affection and education. The love poem elevated the married woman above the previous conception of her as an entity whose value was defined by the capacity to spawn male offspring. While individualism is typically associated with capitalist society, there were forms of individualism which developed in court society during the Renaissance and these forms of self-consciousness were incompatible with the social rigidity imposed by the patriarchal household.

The cultivation of the self-conscious individual was an effect of growing mercantile urbanism, the troubadour conception of individualized emotions, university culture and the autonomy of city life (Chenu, 1969). With the emergence of Renaissance culture in the fifteenth century, powerful women within the court circles of the major Italian towns began to assert greater personal autonomy and individuality against traditional patriarchal control (Heller, 1978). In his classic study of the Renaissance, Burckhardt (1960) also wrote about the growing eminence and individuality of women in the context of aristocratic Italy. Women played a major role in setting the cultural tone and ambiance; they became respected for their learning and culture as equals alongside men. It is important, however, not to overstate this argument about the freedoms of Renaissance women. The argument applies to a small elite of women within the court, and the equality enjoyed by them has to be set against the background of marriages which were contractual and economic rather than companionate and affectionate. Indeed, part of the 'freedom' enjoyed by these women to engage in romantic and adulterous alliances was explained by the restraints of the conventional marriage:

> After the briefest acquaintance with her future husband, the young wife quits the convent or the paternal roof to enter upon a world in which her character begins to develop rapidly. The rights of the husband are for this reason conditional, and even the man who regards them in the light of a 'ius quaesitum' thinks only of the outward conditions of the contract, not of the affections. (Burckhardt, 1960, p. 270)

Although there was considerable sexual licence in such a context, a woman who lost her 'honour' in a flagrant and public way was often subject to violent and bloody punishment, as was her lover. The 'honour' of the family was of vital interest not only to husbands but also to brothers and uncles. Hence the 'freedom' of these elite women was hedged around by threats of physical violence and social stigma, since honour was not an elastic commodity.

It is often held that the Reformation shattered the traditional Catholic, mediaeval world in which the family was based on an economic contract and affectionate love was to be discovered outside and illicitly. The Reformation did not make the body clean, but it did in many respects transform the nature of family life and sexuality. The Puritan revolution dismissed the confessor from his traditional position of authority within the family circle, making the husband pre-eminently responsible for the nature of everyday family life. At the same time, celibacy and the priestly calling became less important in the scale of religious values, as domesticity itself became a lay vocation. Religious education became increasingly domesticated as parents assumed duties previously dominated by ecclesiastical authorities (Zaretsky, 1976). Thus, the Puritan revolution 'by reducing the authority of the priest in society, simultaneously elevated the authority of lay heads of households' (Hill, 1964, p. 43). However, puritanism had very paradoxical consequences for both women and children. Children were regarded as unreformed bearers of original sin, who therefore required intensive indoctrination and meticulous supervision. The Calvinistic ideal embraced the notions of ingrained infantile depravity and juvenile frivolity, mitigated by instant conversions and precocious saints (Grylls, 1978). The adolescent body became highly charged with strong forces of damnation and salvation, constrained by the power of individual conscience under the prompting of the church. Fathers had a new incentive to secure the compliance of their children against the overwhelming facts of original sinfulness. The aim of childtraining was to break the recalcitrant will of the infant (L. Stone, 1979).

In the same period, while women gained a certain religious status within the household, they also lost a certain degree of control over domestic arrangements, such as the celebration of traditional religious festivals within the home. There appear to be three reasons for the decline of status for wives in the sixteenth and early seventeenth centuries. The diminution of the importance of kinship left the wife more at the mercy of her husband within the home; the decline of Catholicism meant that women could no longer appeal to the cult of Mary as an expression of the religious significance of women; legal changes reduced the rights of wives in relation to domestic property (L. Stone, 1979).

## Witchcraft

The Protestant Reformation also had the unintended consequence of exposing women more profoundly to witchcraft accusations. Protestantism removed much of the protective veil of ritual and magical practices which had protected the mediaeval lay person from evil and witchcraft. Puritanism proscribed the use of holy water and holy wells, denied much of the instrumental effectivity of prayer, diminished the importance of sacraments, frowned upon the churching of women and abolished the semi-magical practices of a previous era (Thomas, 1971). The result was that the lay person had no protective, counter-magic by which to ward off the threats of witchcraft, evil and the devil. While the Reformation affirmed strenuously the omnipresence of evil, it took away the protection offered by the priest, the sign of the cross, holy water and holy places (Thomas, 1970). The paradoxical result was that from the middle of the sixteenth century evil in the shape of witches appeared to be everywhere (Trevor-Roper, 1967). In particular, women were held to be the principal accomplices with the devil in the crime of witchcraft. Women were regarded as more irrational, emotional and impressionable than men, and hence more susceptible to satanic temptation. This view of women was typically backed up by an appeal to the mythology of Genesis and the fall of Adam, as in the case of the Daemonologie of James Stuart (Later James I) in 1597. It is not surprising, therefore, that in the period from 1563 to 1727 between seventy and ninety per cent of witchcraft suspects in northern Europe were women (McLachlan and Swales, 1980). While women were normally suspected of a pact with the devil involving sexual intercourse, this sexual theme was absent in the case of male suspects. The problem of female witchcraft suggests that men did treat women as pre-social creatures, whose lives were more determined by 'natural' (or 'unnatural') passions than by culture. The nature of the household and property in Puritan Britain still played an important part in the distribution of witchcraft by making old, dependent women more likely to be suspected of witchcraft practices. In Britain, changes in the traditional system of mutual charity at the village level had made dependent women, especially widows, particularly vulnerable to economic hardship. These women became a burden on the community and

> [t]he conflict between resentment and a sense of obligation produced the ambivalence which made it possible for men to turn begging women brusquely from the door and yet to suffer torments of conscience after having done so. This ensuing guilt was fertile ground of witchcraft accusation. (Thomas, 1970, p. 67)

Women who had become detached from the supportive framework of the household unit were thus forced to rely on neighbours rather than their kinfolk for aid and support. Like 'youths' detached from the social structure by their lack of property, these unattached widows were seen to be socially dangerous to the public order. It was not women as such who were accused, but women who had fallen outside the ambit of patriarchal society. The rule seems to be: 'No property, no personality'.

I have argued that the Puritan revolution had very contradictory consequences for both women and children. While family life rather than the church became a more central focus of social concern and while the importance of the family for political life was constantly affirmed, the status of women in society was also to some extent restricted. Although it is possible to detect these shifts in the role of women through various historical periods, the implication of the property argument is that women's social position cannot radically change without a fundamental reorganization of the relationship between male authority, property and household. The question of divorce and the legal rights of women over property are thus crucial considerations in any sociological analysis of patriarchy. In this respect, John Milton's discussion of divorce perfectly illustrates the ambiguities of the Puritan attitude to women, but also anticipates changes in the social location of women with the rise of 'possessive individualism' and industrial capitalism.

By placing a new emphasis on the religious nature of the marriage tie and by treating marriage as a union freely entered into by man and wife, the Reformation also brought to the foreground the whole issue of the dissolubility of marriage. Despite the stress on the importance of freedom of conscience and religious individualism, the Reformation made divorce unacceptable as a solution to an unhappy marriage. By making marriage a vocation equivalent to traditional religious vocations, Puritanism made alternatives to divorce – adultery, homosexuality, prostitution – equally unthinkable. Puritanism, in short, offered no escape from a marriage which was basically Christian but also unhappy. This was precisely the dilemma of John Milton who, at the age of thirty-four years, precipitately married Mary Powell, the sixteen-year-old daughter of a royalist family from Oxfordshire. The couple were emotionally and intellectually incompatible, and Mary Milton soon returned to her parents. Partly as a response to his own experience, Milton published a series of pamphlets on divorce between 1643 and

1645 in which he attempted to present a moral argument in favour of the dissolubility of divorce within a Christian framework.


## An Argument for Divorce

The main point of Milton's defence of divorce was that marriage is not primarily about reproduction but about companionship and where this emotional mutuality fails divorce is a reasonable solution to marriages which have become empty and painful. His main target then was not 'the idealism of Catholic doctrine' but 'the realism of Catholic mediaeval practice' (Grierson, 1956, p. 53) which, in the absence of divorce, covertly accepted adultery as 'a fact of life'. For Milton, the worst condition in life was loneliness and he treated matrimony as one of the basic social provisions for companionship. The union of man and woman was not primarily sexual, but existed 'to comfort and refresh him against the evil of solitary life' (Milton, 1959, p. 235). The idea of marriage as a necessary evil and a concession to the flesh was the construct of mediaeval Catholicism and not a necessary element of Christian theology. Milton did not argue in favour of 'licence and levity' but argued that 'Some conscionable and tender pitty might be had of those who had unwarily in a thing they never practiz'd before, made themselves the bondmen of a luckless and helples matrimony' (p. 240). In the absence of divorce, men would comfort themselves at the stews or in their neighbour's bed. Marriage ought to be based upon the free and amicable communion of minds and thus it is cruelty to force together persons who find, after marriage, that they are incompatible. This cruelty is not unlike 'the custom that some parents and guardians have of forcing marriages' (p. 275). While Milton produced a strong humanistic argument in favour of companionate marriage, his analysis of divorce proved to be difficult to reconcile with the Reformation view of marriage as a life-long commitment. As a result, Milton came increasingly into conflict with Parliament and the Presbyterians, and was forced to defend other aspects of his position, such as the importance of freedom of communication in the *Areopagitica* of 1644.

Milton's view on marriage and divorce pinpointed the contradictions within Puritanism as a whole. Milton, along with other Puritans, was committed to the importance of filial duty and parental authority, while also believing that parents had no right to select marriage partners for their offspring. Milton had a profound commitment to individual freedom and the liberty of conscience, while also believing that the state and the church should be responsible for education and the moral development of the individual. Milton held that the primary justification for marriage was intellectual companionship, while also believing that it was the right of the husband to decide whether the marriage was providing companionship. Divorce was to exist as a method by which husbands, who found their wives no longer acceptable as companions, could dissolve marriage. Milton presupposed much of the traditional framework of gerontocracy and patriarchy, while also pressing forward an argument which presupposed individualism. The antinomies in Milton's thought thus reflected the disjunctures of the period. The first half of the seventeenth century saw the statement of a clear political theory connecting the absolute right of kings with the absolute right of husbands within the patriarchal household, but this period also saw the origins of social contract theory and individualism which were incompatible with the theory of patriarchal powers. Milton's view of divorce, while being in many respects limited and traditional, represented a turning point in the relationship between power, property and patriarchy.

In this chapter I have examined the broad religious background to patriarchal ideology which regards women as by nature emotional, irrational and unstable. This religious view suggests that women's natural passions are more potent than their powers of reason: Eve's body governs Eve's mind. The history of Christian attitudes towards women is thus powerful evidence of the validity of the feminist argument that women are subordinated in society by an ideology which treats women as closer to nature than to culture. What supports this patriarchal ideology is, however, the control of property within the household so that, in practice, it is difficult to separate patriarchy and gerontocracy. Women were subordinated, but they were subordinated alongside other dependents of both sexes. The ideological control of women and 'youth' is grounded in the control of property. Once household property becomes less essential to the maintenance of society, divorce becomes possible and readily available. Changes in the nature of property and the household are thus the material basis for changes in social relations between men and women. Patriarchy as a result is weakened and transformed into 'patrism'.

# 6

# From Patriarchy to Patrism

## The Dominant Ideology Thesis

Marxism is often taken to argue that the institutions, culture and social practices of a society stand in a necessary or determinant relationship to the dominant mode of production. This position could at least be taken as an unobjectionable implication of a strict interpretation of historical materialism. It is well known, however, that in practice it has proved notoriously difficult to demonstrate both theoretically and empirically that any particular ideology, such as 'individualism', stands in a necessary connection with the capitalist mode of production and even more difficult to demonstrate that an ideology is dominant in capitalist society. *The Dominant Ideology Thesis* (Abercrombie, Hill and Turner, 1980) has been criticized on a number of grounds, not least of which is that the book failed to consider sexism and patriarchy, among others, as playing a crucial ideological function in capitalism. In this chapter, it is argued that capitalism, far from requiring patriarchal domination, actually undermines patriarchal power by transforming the nature of the household unit. Insofar as patriarchy survives, it is largely a defensive ideological reaction against socio-economic changes which erode male dominance in both the public and private spheres. An implicit theme of this presentation is that the transition from early to late capitalism is more important in the transformation of the household unit than the transition from feudalism to capitalism (Turner, 1981). One further consequence of this assessment of the theory of patriarchy is that it would be difficult to claim that both patriarchal and individualist ideologies are dominant within capitalism, since it is shown that these are, in fact, mutually incompatible.

The concept of 'patriarchy' has become central to feminist theory, especially since the publication of Kate Millett's *Sexual Politics* (1977) in 1969, where the notion of patriarchal power played a major analytical role. The analysis of patriarchy is clearly important for the development of feminist politics, but it equally raises major issues for both Marxism and sociology. At the nub of these issues lies the question of economic determinism. Defined simply as the domination and subordination of women by men, patriarchy as a power structure appears to exist under a variety of modes of production – slavery, feudalism and capitalism. While it has a specific relationship to the household and to capitalism, it exists outside the household and often persists despite changes in the form and function of the household. Patriarchy pre-dates capitalism and the theory of patriarchal authority can be found in the political writings of Plato and Aristotle, but it is also present in Locke, Rousseau and Mill (Okin, 1980). In Western culture, patriarchal authority found clear ideological support in the Christian view of women as morally evil, but it is equally present in so-called oriental religions. The very ubiquity of patriarchalism in social time and space makes the possibility of a definite detailed causal explanation appear remote. In this chapter I shall not attempt to 'solve' the analytical puzzle of patriarchy, but I want to suggest that any theoretical conundrum in sociology may be illuminated by a preliminary exercise in the sociology of knowledge, namely by an examination of the social context of the theory itself. It also seems valid to make a distinction between patriarchy as a theory and as a social practice, before attempting to connect these sociologically.

**Patriarchalism**

The idea that the family is the origin of society and its main source of sta-bility made its appearance with the dawn of Western political philosophy. The specific theory of patriarchal power, however, is to be located in the English constitutional crisis of the seventeenth century and with the polit-ical writings of Sir Robert Filmer who died in 1653 and whose *Patriarcha: A Defence of the Natural Power of Kings against the Unnatural Liberty of the People*, written around 1640, was published posthumously in 1680. It is often argued that Filmer, a member of the Kentish gentry, is remembered in political thought largely because John Locke's *Two Treatises of Government* attacked Filmer's defence of royal absolutism. The general background to Filmerism was the contest between the doctrine of the divine right of kings as set forth, for example, by James Stuart in the *Trew Law of Free Monarchies* in 1598 and the contractualist theories of writers like Thomas Hobbes in the *Leviathan* of 1651. In his study of Filmerism, Gordon Schochet (1975) identified three general circumstances which contributed to the emergence of a specific patriarchal theory of political authority.

One immediate cause of the rise of an articulate patriarchal theory was the ideological threat posed by contractualism. Patriarchy was the tradi-tional political world-view but before the seventeenth century this was largely taken for granted. The Fifth Commandment – honour thy parents – was part of the standard catechism for children in the Church of England and in politics it was assumed that the family was the root of all secondary social institutions since it was the family which provided the institutional linkage between the individual and the public sphere. States and empires were simply enlargements of the household in which they had their histor-ical origins. Political institutions were natural, not man-made conventions, since political life had its inception within the household which was a nat-ural environment. The articulation of patriarchalism can thus be seen as partly a defensive reaction to alternative contractualist theories which made a clearer distinction between social and political institutions and which regarded human institutions, especially political ones, as conventions and therefore capable of change. A second aspect of the growth of Filmerism was a new conception of the family which grew out of the English Reformation. Celibacy ceased to be the norm of a true religious vocation and there was a greater theological emphasis on the religious values and duties of family life, especially fatherhood. Marriage and the rearing of children were more highly valued by contrast with virginity. Fatherhood became a vocation, replacing the fatherhood of the Catholic priest. In addition, fathers within the household now had absolute power since they were no longer under the control of the priest. The parallel between domestic and political absolutism now became obvious. The political doctrine of patriarchalism actually fitted the daily experience of typical Elizabethan and Stuart households. The so-called 'extended family' of the period typically consisted of a married couple, their unmarried children, domestic servants and apprentices. Given these demographic characteristics, the presence of grandparents in the household was unusual (Laslett, 1968). The family was essentially a two-generation unit, but it had a group of domestic servants living under the same roof. In other words, the patriarchal authority of the head of the household extended over a heterogeneous collection of individuals, both male and female. Subordinate male domestics were thus subject to patriarchy as much as the patriarch's wife and children. The household as a rigid, authoritarian structure matched the hierarchical organization of the public realm, making the analogy between domestic and political fatherhood highly plausible. Finally, Schochet argues that the dominance of what he calls a genetic theory of politics against a teleological view made it possible to derive political obligation from an account of the origins of society. The evolution of society does not, from a genetic point of view, provide norms for political action, since the norms of a political order can only be discovered in its origins. This antihistorical view of the political community made it possible for writers like Filmer to establish the basis of the authority of the Stuart monarchy on biblical patriarchy, especially the absolute authority of Adam.

As a political doctrine, patriarchalism can be said to have three components. There is first anthropological patriarchalism which attempted to provide a description of how the family was expanded by population growth, territorial extension and conquest to become a state. In other words, a description of how patriarchalism becomes patrimonialism. When this description was combined with a genetic theory, it was possible to link this distinction to some moral doctrine that the origins of an institution were authoritative with respect to future commitments. Geneticism made it possible to move from 'is' to 'ought', from description to moral evaluation. The second component of patriarchalism, moral patriarchy, follows directly from this transition from description to evaluation. Political obedience was justified on the ground that political authority had originally belonged to fathers and this argument was normally supported by the evidence of the Old Testament. Divine right absolutism was thus sanctioned on the grounds that it was the first form of government in the Genesis account of the origins of human society. Thirdly, there is what Schochet simply calls 'ideological patriarchalism' which employed the fatherly image as the basis for all forms of authority, including kings, fathers, magistrates, teachers and masters. Ideological patriarchalism was not particularly dependent on a genetic theory of origins. The argument was simply that God had commanded obedience to fathers and by extension to kings who were fathers of the realm.

Filmer's doctrine of patriarchy was in two sections – a critique of Hobbesianism, especially the idea of a state of nature and a social contract, and a positive affirmation of all three components of the patriarchal theory of authority. Filmer's logical argument can be summarized as follows:

1   Familial authority is natural, divinely sanctioned, and – in its pristine form – absolute and unlimited.
2   Political power is identical to the power of fathers. Therefore, political power is natural, divinely sanctioned, and – because it enjoys the ancient and original rights of fatherhood – absolute and unlimited. (Schochet, 1975, p. 269)

While Filmerism was obviously appealing as a theory of household and political absolutism, it was equally widely attacked by Locke, Rousseau and the Scottish Enlightenment philosophers. Locke emphasized the reciprocity not hierarchy of domestic obligations, rejected absolutism and argued for limited powers. For Rousseau, there was a major distinction to be made between the private and the public economy; the metaphor of fatherly powers provided no basis of kingly authority. For the Scottish commonsense philosophers, historical research emphasized the relativity of human arrangements and there was, for writers like Hume, an unbridgeable gap between 'is' and 'ought'.

Schochet argues that patriarchalism as a meaningful political ideology went into rapid decline after 1690. The causes of this decline were two-fold. First, the restoration of the Stuart monarchy and the revolution of 1688 made the claims of the theory of divine right of kings factually difficult to support. Contractualism appeared to fit the facts of political life more directly and neatly. Secondly, the dominant political doctrine from Locke onwards was individualism and with it a bundle of political beliefs – separation of powers, limited government and private property rights – which were incompatible with absolutism. Individualism as a political doctrine which invests rights in persons *qua* persons thus appears to be incompatible with patriarchy which invests rights in men *qua* fathers. It is well known of course that Locke's individualism was possessive (Macpherson, 1962) and invested rights in persons who already enjoyed property rights. One solution for liberal individualism in practice was to regard women as not persons, with the result that liberal philosophers like Mill remained consistently inconsistent with respect to the rights of women and children. Prior to the Matrimonial Causes

Acts of the late nineteenth century, on entry into marriage women were no longer legal persons. Schochet's argument, however, implies that, insofar as individualism is a doctrine specific to capitalism, the growth of capitalist individualism ought in theory to be incompatible with the continuity of patriarchy. The implication is that patriarchy is a traditional theory of authority which would appear archaic in a society characterized by individualism, secular values and legal-rationalism. For example, to take a theory like that of Lucien Goldmann, the growth of exchange relationships under capitalism creates a system of values centred on the individual, freedom and universalism (Goldmann, 1973). In principle, we would except a society based on commodity-exchange between individuals to limit or undermine patriarchy based on a system of the exchange of use-values within the household. I return to this point shortly.

Filmerism is obviously a primitive if not vulgar form of patriarchy, but a study of Filmerism is instructive for an analysis of the survival of patriarchy in capitalism. While the family had been from Greek philosophy onwards regarded as the principal source of social stability, the specific doctrine of patriarchal authority is the product of a crisis of English political absolutism (P. Anderson, 1974) in the period prior to the bourgeois political revolution of 1688. It was a defensive theory of kingship, but it was specifically part of the Protestant revolution which regarded priests as an interference in the natural rights of fathers. Patriarchy as a theory of male authority emerged during a period of repressive sexual norms, consequent upon the rigidity of Protestant theology with respect to female sexuality (L. Stone, 1979). The political demise of patriarchy as a theory of political obligation corresponds with the rise of industrial capitalism, the dominant ideology of which was supposedly individualism, *laissez-faire* economics and the nightwatchman state. It might be more accurate to say that capitalism required limited individualism, since it excluded women in practice from citizenship. There are nevertheless a variety of social changes which accompanied capitalism – the secularization of society, the decline of natural law theory, the availability of contraception, the decline of the extended family particularly the erosion of domestic servants – which would appear to be incompatible with patriarchalism. In this respect, it would be possible to regard feminism as a movement to establish citizenship for women which followed socialism as an attempt to secure citizenship for working-class men and therefore as a usurpational movement against social closure within the market place (Parkin, 1979). The restrictions on that struggle against social closure would therefore appear to be the topic of modern patriarchy. Before coming to that argument I want to explore a more recent sociological account of patriarchy, namely that presented in *Economy and Society* by Max Weber (1978). One reason for this is that Weber is one of the few classical sociologists to use the term persistently as part of a theory of traditional authority. Furthermore, Weber's sociology raises in an acute form the whole debate about the relative autonomy of economic and political power.

## Weber on Patriarchy

The discussion of patriarchy is located in Weber's formal analysis of the household economy and, more particularly, in his conceptualization of domination, where patriarchalism is the 'pure type' of traditional domination. Patriarchy is defined simply as the personal power of a master over his subjects (wife, children and servants) within the household. The authority of the patriarch is grounded in the norm of filial piety, reinforced by propinquity and the daily routines of common life. Patriarchy emerges out of and creates interpersonal dependency:

> The woman is dependent because of the normal superiority of the physical and intellectual energies of the male, and the child because of his objective helplessness, the grown-up because of habituation, the persistent influence of education and the effect of firmly rooted memories from childhood and adolescence, and the servant because from childhood on the facts of life have taught him that he lacks protection outside the master's power sphere and that he must submit to him to gain that protection. (Weber, 1978, vol. 2, p. 1007)

The authority of the master is customary and traditional, and these norms are typically backed up by an appeal to the sacred. However, it is interesting to note that his daily authority is located in a variety of sources – the alleged weakness of women and children, habituation, ideology and the powerlessness of the servant outside the household. Furthermore, patriarchal power is limited and uncertain, rather than absolute and unquestioned.

The patriarch's power is checked by custom and tradition. It is in principle difficult for the patriarch to appeal to tradition as the basis of his own authority while also systematically overriding tradition in order to exploit openly his subjects. A second source of political weakness lies in the simple fact that the patriarch is greatly outnumbered by his dependents who may collectively seek to oppose his power. In addition, when the family or household is extended by natural expansion or conquest, patriarchal power is converted into patrimonialism which Weber defines as 'a special case of patriarchal domination – domestic authority decentralized through assignment of land and sometimes of equipment to sons of the house or other dependents' (vol. 2, p. 1010). Because of the extended nature of this control through decentralization, reciprocal rather than dependent relations become central, which results in a further limitation on direct, unlimited power of the original patriarch. The picture that emerges from Weber's account of patriarchy is not that it is permanent and unchallenged, but rather that the master is caught in an uncertain and potentially conflictual relationship with his subordinates.

What is also obvious from Weber's formal sociology is that patriarchy and patrimonialism are pre-modern forms of domination which were characteristic of traditional societies before the emergence of legal-rational authority with capitalism. Weber argues that the sexual relationship between men and women can only become the basis for social action when it is transformed into a specific economic organization, namely the household. This unit came into existence for the organized and collective cultivation of soil. The household in this sense is not, for example, characteristic of nomadism. The household which is based on patriarchal authority exists for the satisfaction of its members and defence against the outside social order. It implies a certain solidarity and household communism for the consumption of goods and services, and this solidarity ruled out inheritance on an individual basis. The maintenance of household solidarity also required sexual regulation of its members 'in the interest of safeguarding solidarity and domestic peace' (vol. 1, p. 364). Sexual regulation led to a situation where heads of households held exclusive sexual rights over their women, although Weber notes that the exclusivity of sexual relations is highly precarious. In short, the household as an economic unit existed for collective production and reproduction, and Weber derives most of the features of the household, such as patriarchy, traditional authority, decentralization and sexual exclusiveness from this central fact of economic survival.

Patriarchy, we might argue, was the political structure which sat logically on the economic base of the household. It follows that the dissolution of the traditional household undermines the economic base which supported the patriarchal superstructure. This argument appears to be the one taken by Weber in his analysis of the impact of capitalism on the traditional domestic economy. The division of labour and the rise of individualism tend to break up the original solidarity of the household as the individual comes to 'shape his life as an individual and to enjoy the fruits of his own abilities and labor as he himself wishes' (vol. 1, p. 375) The individual in capitalism relies less on the household and kin for protection because the state assumes these functions. The household and workplace become ecologically separated and the household ceases to be a unit of common production becoming instead a unit of common consumption:

> Moreover the individual receives his entire education increasingly from outside his home and by means which are supplied by various enterprises: schools, bookstores, theatres, concert halls, clubs, meetings, etc. He can no longer regard the household as the bearer of those cultural values in whose service he places himself. (vol. 1, p. 375)

The household is no longer a unit of production and is replaced by the firm, which is increasingly detached from any kinship basis and operates according to bureaucratic, legal-rational norms rather than by tradition. The implication of this account of the transformation of the pre-modern household must be that patriarchy is weakened by the decline in the size of the household, the rise of individualistic values, the transformation of traditional authority and the disenchantment of religious values. One other implication is, however, that the demise of patriarchy is limited by the fact that the household remained a unit for consumption and reproduction with women excluded from the public domain by their allocation to the functions of reproduction. How this exclusion could be legitimated in the absence of general values (either of a religious or natural law variety) remains obscure and Weber has nothing to say on this issue.

There are a number of general comments to be made on Weber's theory of patriarchal power. First, Weber associated patriarchy with gerontocracy as a prime illustration of traditional authority or more strictly domination, and therefore it is evident that patriarchal power is pre-modern. Secondly, the discussion of patriarchy is merely a prelude to a more central concern of Weber's, namely the contrast between patrimonialism and feudalism in relation to the transition to capitalism. Weber's notion that the patrimonial state is an extended and enlarged version of the patriarchal household appears to resemble the Filmeristic anthropological description of how one gets from fatherhood to kingship. Thirdly, in contrast to Filmerism, Weber does not want in principle to derive any normative conclusions from either the origins or transformation of patriarchalism, although in practice his analysis does contain many evaluative assumptions – such as the natural superiority of the physical and intellectual energies of men over women. Fourthly, patriarchy is a relationship of domination over both women *and* men; male servants are subordinated to the master alongside subservient women and children. Fifthly, Weber's sociology is organized round two questions (ownership of the means of violence/ownership of the means of production); the relationship between violence (which is the real focus of political power) and production of economic goods is variable. However, in his account of the household, it appears that the economic relations of production determine the form of politics. For example, the transformation of the household by the external forces of capitalist development must undermine the traditional authority of the master/patriarch. Since the household is no longer a productive unit, the importance of conserving familial wealth for future investment becomes less significant with the development of banking, long-term credit and the autonomy of the company from kin control.

**Engels on Patriarchy**

It is interesting to contrast Weber's view with that of Engels in *The Origin of the Family, Private Property and the State*. In this study of the family, Engels presented an unambiguously evolutionary view of the development of the family: under savagery, there was group marriage; under barbarism, pairing marriage; for civilization, that is capitalism, there was monogamy, supplemented by prostitution and adultery. Monogamy, primogeniture and the subordination of women all correspond to the emergence of private property in capitalism. The need to control women under a system of patriarchy is an effect of the need to control property under primogeniture (Abercrombie, Hill and Turner, 1980). For Engels, patriarchy is not undermined or reinforced by capitalist relations. However, Engels was able to add to this position an analysis of class which was largely absent from Weber. Under capitalism, the function of the working-class household is the reproduction of labour; the function of the capitalist household is the reproduction of capital. However, Engels also recognized that the growth of Protestant individualism undermined the idea of the family as an economic contract and created a space within which women could experience some liberation of individuality. Once again we come up against the argument that insofar as capitalism fosters individualism, it generates a doctrine that it is, at least in principle, incompatible with the traditional claims of patriarchal authority. The idea of romantic love as the basis of the choice of marriage partners within an individualistic culture would appear to be incompatible with the idea that patriarchal preference should be the basis of marriage. Engels solved this by arguing that paradoxically the bourgeoisie was still committed to the antique doctrine that choice was not a suitable basis for marriage, while the proletariat had in practice adhered to the notion of individual choice on the basis of affection as the criterion of a marital relationship. Despite the differences between Marxism and Weberian sociology, there is one common element – namely that individualism is corrosive of patriarchal absolutism and that the former is the product of capitalism. Unfortunately it is very difficult to give an exact periodization of individualism (Turner, 1983). For those writers who locate individualism in Renaissance culture, the autonomy of women from patriarchal structures is to be located in the autonomy given to aristocratic women within the aristocratic courts of Renaissance Europe (Heller, 1978). Despite these difficulties of interpretation, there is some agreement that in many respects capitalism and patriarchy are incompatible. The decline of the household as a productive unit, the growth of individualism, the emergence of universalistic values and the division of labour should undermine traditional patriarchy as a dominant theme of traditional power. At least formally, capitalism is not based on ascriptive values (Parsons, 1951). Just as city air made the peasant free, so in theory capitalism should undermine the traditional domination of women by heads of households.

Weber's contribution to the contemporary feminist debate about patriarchy thus appears to be the argument that (1) patriarchy is rooted in pre-modern economic structures of the agrarian household, and (2) capitalism begins to undermine patriarchy by converting the household into a unit of consumption via the ideological agency of individualism. There is, however, an equally important Weberian version of patriarchy which comes from his concept of social closure. By this term, Weber meant any practice which was aimed at the monopolization of social rewards by a group of eligibles against outsiders. Such exclusionary practices could be based on almost any criterion – age, sex, ethnicity, religion, language, nationality or colour. Weber suggested that almost any human characteristic could be seized upon as a suitable basis for defining exclusion; the criteria for the exercise of domination would thus appear to be entirely arbitrary. Such a theoretical position seems untenable and Parkin (1979) has suggested that exclusionary criteria are never arbitrary. Exclusionary criteria emerge because the state has already given some sanction to their selection. For example, he argues that ethnicity as a criterion of exclusion arises because ethnic groups have already suffered disabilities through conquest or forced migration and their inferior status has already been acknowledged by law. In the case of women, he suggests that their current exclusion is an effect of historical forces, namely their second-class status in law.

I have attempted to identify two possible views of the position of women in society from a Weberian perspective, namely patriarchy and social closure. It seems to me that in Weber's sociology it would be possible to see these two in historical sequence. First, capitalism weakens patriarchy by transforming the household as an economic unit, but secondly social exclusion of women continues because the long history of patriarchy backed up by state legislation of matrimony provides the foundation for the contemporary subordination of women.

## Feminist Theory

On the face of it, the feminist analysis of patriarchy looks very different from that of Weber. The feminist argument is that the development of capitalism did indeed separate the home and the economy as the household ceased to be a unit of production. Capitalism, however, maintains patriarchy for three interconnected reasons (Kuhn and Wolpe, 1978; Rosaldo and Lamphere, 1974). Firstly, it is in the interests of capital to support the unity of the family because the household is the primary unit of consumption. For Filmer, the unity of the family was primarily of political significance; for feminists, it is the location of the family within the circuit of commodities which appears essential. Secondly, women are locked in the private space of the household because women continue to have crucial reproductive functions for capitalism in providing fresh labour. This reproductive function is reinforced by patriarchal ideologies which claim that women, to quote Weber, lack the physical and intellectual energies of men. It is thus 'natural' for women to stay at home and have children because they are crucial for the mothering and nurturing of men. Thirdly, women cheapen the cost of labour for capital since their servicing of men is unpaid labour. Patriarchy is thus an effect of capitalism while also being a condition of its existence.

The argument I wish to present is that the model of the household which lies behind this analysis is relevant to early but not to late capitalism. The model assumes that (1) the household is a unit of consumption in which women produce use-values which are collectively consumed by the family, (2) women are largely and exclusively relegated to the domestic sphere, (3) bourgeois women reproduce eligible heirs and are therefore subordinated by various legal and religious ideologies, and (4) working-class women reproduce labour. My argument is simply that these features are not necessary requirements of capitalism, but historically contingent and variable. It follows that patriarchy is not a necessary requirement of capitalism and that Weber is at least partly correct in arguing that some aspects of capitalism actually undermine patriarchy. Furthermore, patriarchy is now a defensive or reactive ideology which seeks to maintain a socio-economic situation which has been significantly eroded. Much of the feminist argument appears to hinge on the reproductive role of women within the family, but this is a one-sided and static characterization of women's position. It is obvious that one peculiarity of women's social location is their contradictory role as domestic labourers and workers in the economy: they produce both use-values and exchange-values when they enter the external labour force (Gardiner, 1975). The crisis of the First World War and the economic boom of the 1950s and 1960s created new demands for labour and drew women into the employment market in large numbers. In Britain, for example, the percentage of married women in the workforce increased dramatically from ten per cent in 1921 to forty-nine per cent in 1976. Much of the radicalism of twentieth-century feminism can be traced back to this experience of female industrialization. Women in Britain came to play an important part in a series of industrial disputes as a result of which they became increasingly aware of the ineffectual role of male-dominated unions in promoting equality between men and women in the work-place. By industrializing women, capitalism also radicalized them. Two further comments can be attached to this rather obvious point. At the same time that women were entering the workforce in the post-war period, most capitalist societies were also increasingly dependent on guest-workers from the colonial periphery. In the 1970s, fifteen per cent of the manual labour force in Britain and Germany were migrants; in France, Switzerland and Belgium, twenty-five per cent of the industrial work force were foreign workers. The position of the guest-workers is not unlike that of women as a whole. They cheapen the cost of labour because they are produced outside the core capitalist states, so that the periphery bears the cost of reproduction. Furthermore, since they occupy typically unskilled or low status positions in the economy, and often enter on a contract basis, they can be returned to the periphery, thus reducing the cost of welfare, unemployment and other benefits. The extent of the use of migrant, especially contract workers, does raise the question of whether capitalism is so dependent on indigenous women as reproducers of labour. At least in principle, capitalism could operate on the basis of a very low internal fertility rate and relatively high rates of migrant labour inputs. Such a situation in which the periphery makes a major contribution to labour replacement was characteristic of the 1970s in Europe. Given inflation and stagnation, capitalism appears to be moving to a situation of permanent high unemployment, rapid advances in key economic areas based on modern technology which are not labour intensive, low fertility and declining use of guest-workers on a contract basis, combined with the export of technology to the periphery where wage-costs remain low. In terms of employment the decline in heavy industry, typically dominated by men, especially ship-building, mining and the car industry, means that working-class men are being rapidly returned to the unproductive domestic sphere. The indications are that men will be increasingly confined to households in which the structural bases of patriarchy are, if not undermined, at least threatened. Where patriarchal values protect them from domestic labour, they will neither produce use-values nor exchange-values.

## The Household in Capitalism

In addition to these changes in the nature of employment, the whole character of the family unit in late capitalism has changed fundamentally. The statistics published by the American population census of 1980 give some idea of the speed of change within the household unit over a period of ten years. While the census reported that 97.5 per cent of Americans live in households, this category is very heterogeneous. Family households are defined as two or more persons living together and related by birth or marriage. Non-family households include two or more unrelated persons, but this also includes 'single-person households' of the non-familial variety. Between 1970 and 1980, the number of non-family households increased from 18.8 to 26.1 per cent. The average size of all households declined from 3.14 to 2.75 persons. While the number of married couples rose by only 7.7 per cent, families with unmarried heads increased by 52.3 per cent. Non-familial households were broken down in the following categories:

| | |
|---|---|
| Two men living together | 25.1% |
| Two women living together | 19.2% |
| Mixed sex units | 55.8% |

In America while there were some fifteen million new households created between 1970 and 1980, only 22.0 per cent of that growth came from couples who were married. Somewhat similar figures could be given for other societies. In Australia, where there has historically been an imbalance in the sex ratio producing a relatively low rate of marriage, in the 1980s one-third of all households was defined as a 'singles household', that is the never married, *de facto* marriages, widowed, divorced and separated (Staples, 1982). Between 1969 and 1981, the number of one-parent families doubled and by 1981 almost 13 per cent of families with children were one-parent families (Harper, 1982). On the basis of his comparative and historical study of the family, Shorter (1977) argued that the modern family is characterized by a widening generation gap, increased instability and the demolition of the 'nest notion' of nuclear family life. He concluded:

> The nuclear family is crumbling – to be replaced I think, by the free-floating couple, a marital dyad subject to dramatic fissions and fusions, and without the orbiting satellites of pubertal children, close friends, or neighbours . . . just the relatives, hovering in the background, friendly smiles on their faces. (Shorter, 1977, p. 273)

I argued that the rise of Filmerism was a defensive reaction against the bourgeois theory of the social contract and the political rights of individuals in favour of the 'natural' authority of kings as fathers. While Filmerism was made increasingly less credible after 1688 in the new climate of individualism, patriarchy as such could survive on the economic back of the traditional household. From both a Weberian and a Marxist perspective, however, patriarchy would be undermined by the impact of capitalist relations on the unity of traditional family life. In early capitalism, the family was transformed into a unit of consumption and came to dominate the private domain of emotion, leisure and privacy. Since women were progressively excluded from the market place, their inferior status in society was reinforced by their reproductive functions. As women began in the twentieth century to move back into the labour market, they suffered a double exploitation as unpaid producers of use-values in the home and low-paid producers of exchange-values in the economy. However, within a longer perspective, women became increasingly radicalized as workers (despite the fact that their oppositional ideology was often underdeveloped, fragmented and individualist) since the possibility of organizing women outside the isolation of the home was increased. At the same time, the traditional household was being fundamentally dismantled. Although there are many counter-arguments (for example, the rates of remarriage are still very high), the household unit of parents and two children, with the husband in more or less permanent employment and the wife confined to the home, is rapidly becoming part of the mythological, nostalgic heritage of capitalism.

There is a sort of parallel between seventeenth-century Filmerism and contemporary patriarchy, namely that they are both defensive and backward looking in circumstances where much of the basis for domination in terms of physiological differences has been eroded. Women do, of course, experience second-class citizenship, closure from elite professional positions, everyday sexism and petty discrimination, but they also have much of the legal, political and ideological machinery by which that discrimination can be successfully challenged. Men do, of course, have the experience of preferential treatment in securing house loans, jobs, promotion and other social benefits as men, but they also find themselves persistently outflanked by female employment, anti-discrimination legislation and by feminist ideology which appeals to precisely that bastion of the dominant ideology, namely individualism, which is supposed to give them possessive rights. Patriarchy as ideology is thus a defensive reaction by men in a society where marriage and the marriage contract no longer give them dominance in the household or in the market. Anti-patriarchy is a usurpational tactic of women to remove the historical barrier of patriarchal social closure in societies where individual citizenship is formally regarded as a universal right. The ideology of possessive individualism can be turned against bourgeois patriarchy just as it was against aristocratic Filmerism. As Marx (1926, p. 73) noted:

> The bourgeoisie recognised that all the weapons which had been forged against feudalism could have their points turned against itself; that all the means of education which it had created were rebels against its own civilization. . . . It has become aware that all the so-called civil liberties and instruments of progress were menaces to its own class dominion, which was threatened alike at the social base and at the political apex – that is to say, they had become 'socialistic'.

That is, like the critics of Filmerism, women appeal to the universal rights of citizenship and individualism against the 'natural' rights of fatherhood, but these claims are at least in part successful because the economic roots of patriarchy in the traditional household have been contaminated by capitalism itself. What remains of patriarchy is a mere vestige of power, an arcane contingency on the outer shell of capitalist society. Capitalism produces patriarchalism by reaping the advantages of cheap labour and unpaid domestic services within the household; it also destroys patriarchy by creating, at least formally, universalistic values and individualism, and, through the demand for labour, draws women into the labour force, radicalizing their consciousness and undermining the nuclear family as an emotional nest. As Marx persistently commented, capitalism is the most revolutionary and progressive movement in human history, because it is forced to revolutionize constantly the whole economic basis of society. This paradoxically radical nature of capitalism partly explains why it is that patriarchalism is part of the world we have lost:

> Every relationship in our world which can be seen to affect our economic life is open to change, is expected indeed to change of itself, or if it does not, to be changed, made better, by an omnicompetent authority. This makes for a less stable social world, though it is only one of the features of our society which impels us all in that direction. All industrial societies, we may suppose, are far less stable than their predecessors. They lack the extraordinarily cohesive influence which familial relationships carry with them, that power of reconciling the frustrated and discontented by emotional means. Social revolution, meaning an irreversible changing of the pattern of social relationships, never happened in traditional, patriarchal, preindustrial human society. (Laslett, 1968, p. 4)

Laslett goes on to note, however, that precisely this closeness and hierarchy of the Stuart household produced the interpersonal social violence portrayed in the dramas of Shakespeare. One indication of the gap between our world of psychologically mobile narcissistic personalities and the structurally rigid dramas of Shakespeare's society may be illustrated by a comparison between *King Lear* or *The Tempest* and the fluid complexity of Eliot, Pinter and Beckett.

Women still experience sexism in everyday life, but this is a defunct patriarchalism, an interpersonal strategy of dominance on the part of men who find their traditional sources of power increasingly open to doubt. Their sexist patriarchalism is a defensive response of a crisis of identities in a society were machismo values are brought into question by permissive state legislation on homosexuality, children's rights and women's liberation. Genuine gerontocracy along with real patriarchalism is dead for the simple reason that there are no patriarchs. What remains is an intensification of the ideological struggle for citizenship, which results in global unevenness. It may be that the core capitalist states will come to depend on the continuity of patriarchy in the periphery, where, for example, militant Islam has shorn up patriarchalism. Where the organic composition of capital is favourable, these peripheral, patriarchal societies will offer profitable investment, cheap commodities and fresh labour (Turner, 1984). Capitalism thus liquidates patriarchy, only to witness its ideological reconstitution on the outskirts of the global economy.

**Patrism**

In race relations theory (Banton, 1967) it is common to make a distinction between racism as a collection of prejudicial attitudes and racism as a social system in which certain social groups are suppressed and exploited through the operation of the market, political structures and the law. Prejudicial racism may be widespread in a society in which minority or migrant groups are formally protected by law; in principle, institutional racism could exist in a society where in everyday circumstances minority groups were not the target of racist attitudes. In practice, one would expect a system of institutionalized racism to foster racist prejudice in interpersonal relations. Thus, it would be reasonable to argue that societies like Britain and Australia are not characterized by institutional racism because the law and political system are relatively open and universalistic. However, Britain and Australia are racist in the sense that very prejudicial attitudes towards minority ethnic groups are widespread in social relations. I want to suggest that a similar distinction is helpful in the analysis of sexual conflict and gender relations. Institutionalized sexual inequality is traditional patriarchy, whereby on the basis of sexual characteristics people are excluded from economic roles, social status, political power and legal identity. Patriarchy involves the systematic social closure of women from the public sphere by legal, political and economic arrangements which operate in favour of men. As I have suggested, however, it is difficult to separate patriarchy from gerontocracy and thus institutionalized inequality on the basis of sex typically includes the hierarchical structure of age groups. Patriarchy surbordinates women alongside children; indeed in terms of dress and fashion women are often encouraged to adopt childlike garments which distinguish them from the 'adult' attire of men. Patriarchy is an objective social structure which is maintained and constituted by a complex system of legal regulations, political organization and economic arrangements.

Patrism is analytically equivalent to subjective racism, because it involves prejudicial beliefs and practices of men towards women without the systematic backing of law and politics. In terms of taxation arrangements, banking facilities and the availability of credit arrangements, it is no doubt the case that women still experience institutions which approximate to patriarchy, but such inequalities are under attack and are generally regarded as discriminatory and incompatible with existing democratic arrangements. A comprehensive system of institutionalized patriarchy no longer exists in the majority of industrial capitalist societies, where the legal, political, religious and economic restraints on women have been largely dismantled.

The collapse of patriarchy has left behind it widespread patrism which is a culture of discriminatory, prejudicial and paternalistic beliefs about the inferiority of women. The implication of my argument is that patrism is expanding precisely because of the institutional shrinkage of patriarchy, which has left men in a contracting power position. Men as a whole can no longer depend on the law to buttress their dominance within the public and private spheres. Institutionalized patriarchy has crumbled along with the traditional family unit and the patristic attitude of men towards women becomes more prejudicial and defensive precisely because women are now often equipped with a powerful ideological critique of traditional patriarchy. Sexual conflict is now more pronounced as a result of defensive patrism and offensive feminism in a period where the institutionalized supports for the sexual division of labour are in a state of advanced decay.

# 7

# The Disciplines

## Foucault, Language, Desire

In structuralism, much of the recent interest in desire and its negation in discipline finds a location in the work of Michel Foucault, especially in his *The History of Sexuality* (1981). Behind Foucault lies Nietzsche, and the constant references to Nietzsche among the so-called New Philosophers of the right are not occasioned simply by Nietzsche's theory of language, but by his conceptual distinction between Apollo and Dionysus – between the discipline of form and the transformation of ecstasy. In short, contemporary debates in structuralism, especially in the reappraisal of the Freudian conception of the struggle between id and superego, reflect an ancient conflict in Western culture between reason and desire. In this ancient dichotomy, social stability rests on the subordination of the desires of the body by the reasons of the mind, through such institutions as the family, the church and the state. In this view, civilization is abstinence, requiring the denial of the flesh and the control of emotion. The ascetic practices of Protestantism transferred the denials of the monastery into the everyday life of the family, the school and the factory. The history of modern societies can be seen as the rationalization of this ascetic process through various sciences of the body. The work of Foucault makes a massive contribution to our understanding of this process, namely the applications of institutionalized power through rational knowledge to the life of populations and persons. Since Foucault has been important in providing a general framework for this sociology of the body, it is time to confront Foucault head-on, since he is more than a frontispiece. Interpretations of Foucault are legion (Chua, 1981; Kurzweil, 1980; Sturrock, 1979). The aim of this chapter is not to add to the industry of interpretation, but to suggest a particular view of Foucault in order to offer an application to a particular field, namely the science of dietetics.

Foucault derives a variety of positions from Nietzsche. First, there is irony. There is a characteristic argument in Nietzsche that most of the (Christian) values we regard as 'good' – charity, mercy, kindness – are the historical products of 'bad' motives and circumstances – cowardice, weakness, hypocrisy. In *The Anti-Christ*, Nietzsche argued that the Christian virtues were the result of resentment experienced by a slave population as an expression of revolt against dominant manly virtues of strength and courage. Christian virtue was merely suppressed violence. Foucault's work is also grounded in this historic irony. (The sciences which seek to improve the human condition through liberal values – psychiatry, sociology, penology, biology – are themselves inextricably expressions of domination; they seek to know in order to organize.) History is merely the contingent struggle of groups caught up in a will to power: 'The domination of certain men over others leads to the differentiation of values; class domination generates the idea of liberty; and the forceful appropriation of things necessary to survival and the imposition of a duration not intrinsic to them account for the origin of logic' (Foucault, 1977, p. 150). Knowledge, which promises liberty, has its origin in power.

Secondly, Foucault shares with Nietzsche a distrust of the pretence of reason, especially Cartesianism and the positivist sciences such as Darwinistic biology. Descartes suppressed insanity as a necessary requirement for reason itself (Derrida, 1978). Dreams and madness are both forms of error, but they stand in a different relationship to truth. Whether or not I am dreaming, $2 + 2 = 4$, but reason in order to exist must expel the possibility of madness. Men may be mad, but thought itself cannot be. Descartes thus stood at a peculiar conjuncture where the possibility of 'unreasonable Reason' and 'reasonable Unreason' disappeared (Sheridan, 1980, p. 23). For Foucault, the cognitive suppression of madness by Cartesianism corresponds to the institutional suppression and internment of the insane in the asylum (Foucault, 1967). *Ratio* is thus grounded in a particular type of violence and develops through a rationalization of society which involves the detailed, systematic control of the individual.

Thirdly, Foucault's debt to Nietzsche's historical perspective is explicitly documented (Foucault, 1977). Nietzsche's concept of 'genealogy' is, for Foucault, the negation of all teleological, progressive and evolutionary conceptions of history, which regard human society as shaped by the march of humane values, reason and progress. By contrast, Foucauldian history is a series of accidental ruptures, brought about by chance, struggle and contingent conjunctures. In this history, 'Humanity does not gradually progress from combat to combat until it arrives at universal reciprocity, where the rule of law finally replaces warfare; humanity installs each of its violences in a system of rules and thus proceeds from domination to domination' (Foucault, 1977, p. 151). Thus, in *Madness and Civilization* (1967) and *Discipline and Punish*, (1979) Foucault rejected the official histories of psychiatry and criminology whereby reason and freedom finally banished traditional errors and violences committed against the mad and the poor. These new knowledges of the insane involved an extension of institutionalized power, permitting a surveillance of the body through the discourses and practices of psychiatry, criminology and penology. Foucault's reading of history offers no consolations; history is mere chance since

[n]othing in man – not even his body – is sufficiently stable to serve as the basis for self-recognition or for understanding other men. The traditional devices for constructing a comprehensive view of history and for retracing the past as a patient and continuous development must be systematically dismantled. (Foucault, 1977, p. 153)

History is relativized and rationalist perspectivism rejected. What is strange and unfamiliar in human history and society cannot be made familiar by the consoling efforts of historians, whose 'refamiliarization' of history attempts to show that bizarre practices, once correctly understood, are reasonable and hence express the universality of the human essence. There is no unity to the human essence, which is merely the arbitrary construct of language. For Foucault, the unfamiliar is unfamiliar and cannot be domesticated by notions of evolution or universalism. History writing is itself a form of conceptual violence, since it imposes a false uniformity on events, people and places in terms of historical discourse which is, like all language, an arbitrary system of signs.

### The Accumulation of Men

Despite Foucault's explicit rejection of the notion of historical trends and uni-form developments, there is a strong historical theme uniting his various studies of power/knowledge. Despite his hostility to systematic theorizing, there is a 'theory' which offers a causal explanation of the modern world. This theory is primarily focused on the demographic (a theory of popula-tions) and the physiological (a theory of the body). The 'key' to Foucault's own discourse occurs in a contribution to the history of the modern hospital where he explains the transformation of eighteenth-century medicine from charitable aid to a policing of health.

> It arguably concerns the economic-political effects of the accumulation of men. The great eighteenth-century demographic upswing in Western Europe, the necessity for co-ordinating and integrating it into the apparatus of production and the urgency of controlling it with finer and more adequate power mechanisms cause 'popula-tion', with its numerical variables of space and chronology, longevity and health, to emerge not only as a problem but as an object of surveillance, analysis, intervention, modification, etc. . . . Within this set of problems, the 'body' – the body of individ-uals and the body of populations – appears as the bearer of new variables, not merely as between the scarce and the numerous, the submissive and the restive, rich and poor, healthy and sick, strong and weak, but also as between the more or less utilisable, more or less amenable to profitable investment, those with greater or lesser prospects of survival, death and illness, and with more or less capacity for being usefully trained. (Foucault, 1980a, pp. 171–2)

The rationalization of Western society in the late eighteenth and early nine-teenth centuries found a new object of exploration and control – the human body itself. The spread of scientific and techno-rational procedures, having gained a foothold in technology and consciousness, latched onto a new terrain, the body of individuals and the body of populations. The institution-alization of the body in what Foucault calls 'panopticism' made possible a statistics of populations and new practices of quantification in clinical medicine, demography, eugenics, penology, criminology and sociology.

It is population pressure which provides the central motif of Foucault's historical commentaries, because it is this factor which stands behind the expansion and development of new regimes and regimens of control – a profusion of taxonomies, tables, examinations, drills, dressage, chrestomathies, surveys, samples and censuses. The pressure of men in urban space necessitates a new institutional order of prisons, asylums, clinics, factories and schools in which accumulated bodies can be made serviceable and safe. In Foucault's own stylistic conventions, this theme is linked to metaphors of space in knowledge and society – site, domain, position, landscape, terrain, horizon and archipelago. Just as the space of knowledge experiences accumulations of new discourses, so the social space is littered with bodies and the institutions which are designed to control them. Hence Bentham's panopticon scheme becomes the principal model for the detailed organization of political scrutiny over the bodies of both deviants and citizens. The appeal to morality and the deployment of the church are no longer sufficient for the control of individual desire; it becomes necessary to encompass urban populations with new institutions of surveillance and inspection.

Foucault is not a systematic theorist. It does appear, however, that most of his historical examples of disciplines hinge on two crucial events – the social reorganization of France following the Revolution and the social effects of population pressure in the city. The result was a policing of society which in turn was a condition for capitalist expansion; these regulatory controls of population were

> an indispensable element in the development of capitalism; the latter would not have been possible without the controlled insertion of bodies into the machinery of production and the adjustment of the phenomena of population to economic processes. But this was not all it required; it also needed the growth of both these factors, their reinforcement as well as their availability and docility; it had to have methods of power capable of optimizing forces, aptitudes, and life in general without at the same time making them more difficult to govern. (Foucault, 1981, p. 141)

Capital could profit from the accumulation of men and the enlargement of markets only when the health and docility of the population had been made possible by a network of regulations and controls.

The centrality of population pressures to Foucault's analysis of the nature of the modern world can be illustrated by reference to a student of Foucault, Jacques Donzelot, whose *The Policing of Families* brings out specifically the historical demography within the Foucauldian perspective. Essentially what Donzelot seeks to show is that, faced by the crisis of an unregulated urban working class, new alliances were formed in nineteenth-century France between the family and the state, between medicine and the household, and thus between the doctor and the wife. The increase in urban poverty and illegitimacy called for a new social economy, but at the same time 'criticism was aimed at the organization of the body with a view to strictly wasteful use of it through the refinement of methods that made the body into a pure pleasure principle: in other words, what was lacking was an economy of the body' (Donzelot, 1979, pp. 12–13). The regulation of the social body had to be premised on new principles of domestic organization in order to achieve a regulation of the body of individuals. While George Cheyne in Britain was drawing attention to the destruction of the aristocratic body, and hence the personnel of government, by the luxuries made possible by colonial growth, it was also evident in France that the peasantry enjoyed more robust constitutions than their masters, because, despite their poor diet, they had regular exercise. Yet this reserve of able peasant bodies was being corrupted by the move to the city where pauperism converted them into a political threat and undermined their natural healthiness. The solution to these problems involved a reduction of the social costs of reproduction, the location of medical knowledge within the family and the reorganization of the family along guidelines suggested by the domestic sciences. This involved a medicalization of society through the agency of the reconstituted family, namely the weakening of traditional patriarchy by an alliance between mothers and doctors:

> The strategy of familializing the popular strata in the second half of the nineteenth century rested mainly on the women, therefore, and added a number of tools and allies for her use: primary education, instruction in domestic hygiene, the establishment of the workers' garden plots, and Sunday holidays (a family holiday, in contrast to the Monday holiday, which was traditionally taken up with drinking sessions). But the main instrument she received was 'social' housing. In practice, the woman was brought out of the convent so that she would bring the man out of the cabaret. (p. 409)

The family became the site where individuals are formed and trained by the new sciences of the hearth – how to eat, sleep, dress and conduct oneself – and where decentralized political power is to be located for the reform of populations. The family became the locus of rationalization and personal asceticism.

My argument is that, while Foucault rejects any notion of historical logic or teleology, in practice the unifying theme of his work is the rationalization of the body and the rationalization of populations by new combinations of power and knowledge. Furthermore, this rationalization is an effect of population densities which threatened the political order of society in the nineteenth century. There is an important convergence, therefore, between the Weberian perspective on the rationalization of society through the imposition of scientific practices at every point of human life and Foucault's interest in disciplines as the subordination and co-ordination of desire. This convergence can be seen in the discussion of the utilization of medical science to control bodies and populations in Foucault and Donzelot. It is also evident in the relationship between asceticism and capitalism.

### Asceticism

For both Weber and Foucault, religious models of thought and practice provide one historical location for the growth and spread of rational surveillance of human populations. In *The Protestant Ethic and the Spirit of Capitalism*, Weber (1965) argued that the disciplines of the monastery were transferred to the home and the factory, where every believer was expected to exert ascetic control over his daily life. Protestantism meant that, since all domestic and public activities had become a 'calling', the lay person could no longer depend upon the priest to perform vicariously such religious duties. Everyday life came under the scrutiny of individual consciousness and as the individual became liberated from the authority of the priest so he also became more subordinated to detailed regulations. Protestantism destroyed any possibility of depending on magical or ritualistic means of salvation, such as the Christian sacraments, and forced the individual to organize his everyday life in terms of rational pursuits in a secular calling. This was, so to speak, the spiritual origin of a rationalized culture in capitalist society, which came to be reinforced by the factory system, modern forms of bureaucracy and the regulation of the state. For Weber, the disenchanted world of modern capitalism is one which rules out spontaneous and individualistic beliefs and practices, because life becomes increasingly subordinated to bureaucratic plans, to organized codes of behaviour, and to a network of regimens that extend to the most private spheres of life, including our pleasures.

There is an implicit argument in Weber that the regulation of passions under monasticism was, with the Protestant Reformation, reassembled in the Protestant household under the patriarchal authority of the husband, who dismissed, as it were, the priest as confessor of the wife from the intimacy of the hearth. The restraint of passions, especially sexual passions, within the home now came to depend on the exercise of the Protestant consciousness in conjunction with the rigours of the calling in the world. Because 'time is money', idle hands are sinful. Sexuality is pre-eminently wasted time, unless it is geared to reproduction, but this must be achieved without pleasures:

> The sexual asceticism of Puritanism differs only in degree, not in fundamental principle, from that of monasticism: and on account of the Puritan conception of marriage, its practical influence is more far-reaching than that of the latter. For sexual intercourse is permitted, even within marriage, only as the means willed by God for the increase of His glory according to the commandment, 'Be faithful and multiply'. Along with a moderate vegetable diet and cold baths, the same prescription is given for all sexual temptations as is used against religious doubts and a sense of moral unworthiness: 'Work hard in your calling'. (Weber, 1965, pp. 158–9)

In modern parlance, the free play of desire is sublimated in the routines of work. With the secularization of culture, these religious norms of work are gradually replaced by the profane disciplines of the factory with Taylorism and Fordism. A similar argument is to be found in Foucault's account of the rise of disciplined bodies.

Foucault recognizes that the disciplines of the body in Benthamite panopticism were anticipated by the monastery. However, there is an important difference since, while monasticism required a renunciation of the body, the modern disciplines of capitalist society require utility. Panoptic disciplines produce an intensification of aptitudes and abilities, while they also increase subordination by self-controlled mastery of the body (Foucault, 1979, p. 137). This was the new asceticism and it was developed at precisely that point in the history of Western populations when the traditional threat of death from epidemics and plagues began to recede. The accumulation of men and of capital were intimately related; the two processes cannot be separated. The accumulation of men would not have been possible without an increase in capital to sustain that demographic explosion. Similarly, the techniques which developed for the control of useful bodies accelerated the accumulation of capital (Foucault, 1979, p. 221). The new techniques of discipline were essential for the organization of factories. The development of capitalism 'would not have been possible without the controlled insertion of bodies into the machinery of production and the adjustment of the phenomena of population to economic processes' (Foucault, 1981, p. 141). While Foucault does not assume that asceticism produced capitalism, he does recognize that the transformation of monastic into factory disciplines was an important feature of the accumulation of capital. The ascetic practices of individual bodies were thus closely related to the accumulation of men in the eighteenth-century origins of capitalism. In the same manner, Weber argued that a surplus population which can be hired cheaply is a necessary requirement for the development of capitalism (Weber, 1965, p. 61), but he also argued that this population had to be disciplined to make it serviceable.

## Dietary Management

In this chapter, I wish to illustrate these arguments in Weber and Foucault through an analysis of the history of dietetic management. There are two aspects to this illustration. First, diet becomes, with the growth of capitalism, increasingly secularized and this involves, as Foucault suggests, a shift from renunciation to utilization. Secondly, there is a shift from the control of aristocratic bodies to working bodies, which also corresponds to a change in objectives from longevity to utility. While this argument is primarily concerned with the debate concerning disciplines in capitalist society, it has to be recognized that diet is one of the oldest components of any medical regimen. The term 'diet' comes from the Greek 'diaita', meaning a mode of life. As a regulation of life, it has the more specific medical meaning of eating according to prescribed rules. There is a second meaning of 'diet' which is a political assembly of princes for the purpose of legislation and administration. This second meaning comes from the French word 'dies' or 'day', because political diets met on specified days and were thus regulated by a calendar. Diet is either a regulation of the individual body or a regulation of the body politic. The term 'regimen' also has this double implication. It is derived from '*regere*' or 'rule' and, as a medical term, means a therapeutic system, typically involving diet, but regimen is also a system of government as in 'regimentation' or 'regime'. We can see, therefore, that 'diet' and 'regimen' both apply to the government of the body and the government of citizens. This etymological argument further reinforces the argument that metaphors of health and illness are persistent metaphors of social organization (Sontag, 1978).

The medical regimen in classical Greece largely consisted of diet as a mode of regulated living. Diet, of course, was also a recurrent feature of any mode of religious life in Christianity. It was one of the basic elements of asceticism as a regimen for the control of desires. Fasting was a basic element in Roman Catholic asceticism but it also clearly survived in Protestantism. William James (1929) recognized temperance in the use of meat and drink as a basic characteristic of all religious asceticism and, to illustrate its presence in Protestantism, quotes from the life of John Cennick, who was the first lay preacher in Methodism: 'He fasted long and often, and prayed nine times a day. Fancying dry bread too great an indulgence for so great a sinner as himself, he began to feed on potatoes, acorns, crabs, and grass; and often wished that he could live on roots and herbs' (James, 1929, pp. 301–2). Protestantism took monastic renunciation into the world of the lay person; it did not dismantle these traditional religious practices. Ascetic disciplines in both Catholicism and Protestantism were a system of rules of conduct to control the flesh by starvation and renunciation. Asceticism as a term comes from '*asketes*' (monk) and '*askeo*' (exercise): it is a regulated practice or regimen of the body. Although the Reformation in general transferred asceticism from the monastery to the home, there was also an important continuity between Catholic and Protestant disciplinary methods. In order to understand the history of diet for the laity, it is important to state one obvious but important point. Given the uncertainty of food supplies, and given the absence of any great variety in the availability of food for the peasantry through the mediaeval period, early dietaries were primarily directed at the upper classes of landlords and merchants. Although in the twelfth century the reign of King Henry I was a period of expansion in medical knowledge and technique, the king himself died of dietary mismanagement. Ignoring his physician's advice for weight control, Henry feasted on lampreys and died from the resulting ill humours in Normandy in 1135: 'Despite the removal of his intestines, brains, and eyes for separate burial at Rouen and the salting of the rest of his body for transport to entombment at Reading Abbey, his corpse still fouled the air' (Kealey, 1981, p. 117). The regulation of royal bodies was seen to be in the interests of the social body; peasants were expendable. More precisely, food supplies for the peasantry between 1350 and 1550 were adequate, but there was a long-term decline, especially in the production of meat, until the middle of the nineteenth century. Thus, after 'the fifteenth and sixteenth centuries only a few privileged people in Europe ate luxuriously. They consumed huge quantities of rare dishes. What was left went to their servants, and what was left after that was sold to food-dealers, even if it had gone rotten' (Braudel, 1974, p. 1136). The popular dietaries of this period of upper-class luxury were pious tracts directed towards the class threatened by over-eating, gastronomic diversity, obesity and alcoholic poisoning.

Two important figures in the history of dietetics for the aristocracy were Leonard Lessius (1554–1623), the author of the *Hygiasticon, or the Right Course of Preserving Life and Health unto Extreme Old Age*, and Luigi Cornaro (1475–1566), the author of *Trattato della vita sobria* (1558), which was translated by George Herbert in 1634. Diseases are frequently interpreted as manifestations of a deeper malaise in the social structure. Just as cancer is often regarded as a 'disease of civilization' (Inglis, 1981), so obesity in the sixteenth and seventeenth centuries was regarded as a physical manifestation of the social flabbiness of the social system, especially as it impinged upon the life-style of the rich. The disorders to which Cornaro drew attention were the 'bad customs' of the time, namely 'the first, flattery and ceremoniousness; the second, Lutheranism, which some have most preposterously embraced; the third, intemperance' (Cornaro, 1776, p. 14). Cornaro, who was an Italian nobleman from Venice, saw the corruption of Italian cultured society by the Reformation, the falsity of court life and indulgence as leading necessarily to the corruption of the body. The solution to social and physiological pathology was to be sought in the government of the body through diet and discipline. Dieting, especially among the rich, was the main guarantee of health, mental stability and reason. A life founded on temperance and sobriety was the principal defence against the aristocratic affliction of melancholy and the disruptive effects of passion on reason. For Cornaro, therefore, the discipline of diet was formulated within a religious framework as the antidote to the temptations of the flesh. Cornaro and Lessius came to have a long-term significance for the development of a medico-religious discourse concerning the physical, personal and social benefits of dietary management.

In the sixteenth and seventeenth centuries, England was renowned for the melancholic disposition of its inhabitants; a special term was coined – 'the English malady' – to describe the prevalence of this national sickness (Skultans, 1979). The condition was diagnosed by a variety of moralists, but it was Robert Burton's *The Anatomy of Melancholy* that came to provide a dominant and lasting perspective on this national complaint. It is particularly important to note that Burton (1576–1640) was a divine, who, like Cornaro, saw the diseases of the mind as symptomatic of social pathology. One of the central themes of *The Anatomy*, published in 1621, was that idleness was the main cause of social and mental disorder. Burton saw the decay of England in idleness 'by reason of which we have many swarms of rogues, and beggars, thieves, drunkards and discontented persons' (Burton, 1927, p. 49). Burton distinguished between three types of idleness, namely the voluntary idleness of rogues, the idleness that accompanied noble status, and the enforced idleness of the religious. It was the leisurely and idle character of noble existence which particularly concerned Burton and it was to the gentry that his book was specifically addressed, since 'idleness (the badge of gentry)' was 'the nurse of naughtiness' (p. 158). In particular, idleness became the social symbol of wealth, while wealth provided the means of extravagant diets and the absence of labour. The rich 'feed liberally, fare well, want exercise, action, employment (for to work, I say, they may not abide) and company to the desires, and thence their bodies become full of gross humours, wind, crudities; their minds disquieted, dull, heavy' (p. 160).

The combination of leisure and luxury had especially damaging consequences for unmarried women in the gentry class and Burton noted that, while serving women rarely suffered from melancholy, noble women were the principal victims of the English malady. Virginity and nobility both led to idleness and isolation, and hence to melancholy. The cure for this condition was marriage, diet, exercise and religion. When these failed to produce the remedy of unruly desires, Burton recommended 'labour and exercise, strict diet, rigour and threats' (p. 273). The government of female bodies was thus linked via patriarchy with the government of the household. Since Burton saw a great affinity 'betwixt a political and economical body' (p. 63), his dietary was necessarily a political treatise. Society presupposed a hierarchy of political control, descending from the state, through the patriarchal household, to the body and desires.

Many of Burton's anxieties and solutions were reproduced in the following century in the dietetics of George Cheyne, who noted the expansion of trade and the growth of mercantile wealth brought exotic and rich foods into the market place. The result of these civilized luxuries was to 'provoke the Appetites, Senses and Passions in the most exquisite and volumptuous Appetite' (Cheyne, 1733, p. 49). Cheyne's medical discourses were primarily addressed to the urban, idle rich, who were the most exposed to the moral danger of strong drinks and exotic foods. The sedentarized life of London and Bath provided a sharp contrast with the natural vigour of primitive man: 'When Mankind was simple, plain, honest and frugal, there were few or no diseases. Temperence, Exercise, Hunting, Labour and industry kept the Juices Sweet and the Soldis brac'd' (p. 174). In order to reduce the destructive impact of affluence on the digestive system, Cheyne recommended, especially for sedentarized merchants and professional men, a strict diet, regular evacuation, exercise on horseback and 'a Vomit that can work briskly' (Cheyne, 1740, p. xlvii). Cheyne's dietary regimen was intended to subordinate the passions of the urban rich which had been inflamed by excessive consumption of exotic food and drink.

While Cheyne was heavily influenced by Cartesianism and the iatromathematics of the Leyden school of medicine, he regarded diet, exercise and regularity as moral activities which promoted the control of unruly passions. It is, therefore, not surprising that his views were highly congenial to the religious outlook of John Wesley and the early Methodists. Cheyne's dietary regulation was easily incorporated within the Methodist code of ascetic behaviour and Wesley used much of Cheyne's medico-morality as the basis of his own *Primitive Physick* of 1752. Wesley also recommended

Cheyne's *Essay of Health and Long Life* to his mother in 1724, partly because it was 'chiefly directed to studious and sedentary persons'. It can thus be argued that the traditional norm of fasting as an ascetic practice within the monastery was gradually transformed by the Protestant dietaries of Burton, Cheyne and Wesley into a suitable exercise of regulation for the laity, and that the elitist dietetics of Burton eventually reached the working class via the popular views of Wesley and the Methodist chapels. The dietary regimen of the monastery infiltrated the secular household via the Nonconformist chapels and meeting houses. This invasion of dietary management into the home was eventually combined with the broader movement of general hygiene for the working-class family under the auspices of the medical profession. It represented a rationalization and secularization of food which ceased to be a stimulant of desire and became instead, under scientific dietetics, a condition of efficient labour. The vocabulary of passions, desires and humours was replaced by the discourse of calories and proteins. The dietary requirements of specific categories of people became increasingly detailed and rationalized. Whereas writers like Cheyne used very general classifications – the idle, the gentry and sedentary scholar – dietetics now came to analyse the requirements of prisoners, workers, pregnant women, the schoolchild and the athlete. Each illness had its specific diet – pulmonary tuberculosis, diabetes, allergic diseases and rheumatism (Rolleston and Moncrieff, 1939) required individualized dietary regimens. Diets become specific to persons individuated by age, class, sex and condition. My argument is therefore that dietary disciplines of the body in the nineteenth century became progressively individuated, secularized and rational. With this process, the idea of diet as a control of the soul in the subordination of desire gradually disappeared.

## Table Practices

It has been argued that one feature of the incorporation of physiology into society takes place by dietary management of the body. Diet is a cultural practice regulating quantities and types of food for designated categories of person. Both what and how we eat are culturally constituted by practices and beliefs. In discussing dietary changes in society, it is also important therefore to consider the evolution of etiquette alongside the growth of individualized diets within a secular framework. The classic study of courtly table manners is Norbert Elias's *The Civilizing Process* (1978). We can trace the growth of civilization as a transition from a situation where eating is a communal activity with a minimal regulation of individual eating habits to a situation where eating is individualized and hedged around by taken-for-granted norms of correct behaviour. The process of civilization involves the control of emotions and expression through changes in expectations (such as norms regulating spitting at table) and in table equipment (such as the use of spoons and forks). With the rise in civility from the sixteenth century onwards, the communal bowl and the communal meal were gradually replaced by individualized patterns as indicated by the use of individual plates and eating utensils. Manners became more complex and differentiated. In the feudal courts, an elementary code was sufficient – do not blow your nose on the table-cloth, do not return half-eaten meat back to the communal dish, after spitting place your foot on the sputum. Eating practices became increasingly elaborate, specifying what type of fork, knife or spoon was appropriate for what type of dish. Eventually the spittoon and the vomitorium were removed since spitting and vomiting were no longer acceptable at table.

The civilization of the table involved the exercise of collective control over the play of emotions. Elias's discussion of changing attitudes towards knives at table is an important illustration of this phenomenon. In the Middle Ages, the upper stratum of warriors routinely carried knives on their persons and there were few inhibitions on the use of knives during eating. The principal norm was: do not clean your teeth with your knife. The knife is, of course, symbolic of death and pain. With the declining social importance of armed warriors, the knife retained its symbolic significance and there was an increasing regulation of its domestic uses: 'it is the symbolic meaning of the instrument that leads, with the advancing internal pacification of society, to the preponderance of feelings of displeasure at the sight of it, and to the limitation and final exclusion of its use in society' (Elias, 1978, p. 123). Knives at table were not to be pointed, especially at the face, and when passing a knife, the handle must be offered to your neighbour. As table customs became more 'polite', so symbols of emotions and passions had to be controlled or suppressed. With civilization, 'regulation and control of emotions intensifies. The commands and prohibitions surrounding the menacing instrument become ever more numerous and differentiated. Finally, the use of the threatening symbol is limited as far as possible' (p. 125). These changes in manners brought about an enhanced control of the individual by restraints which transformed the affective life of society and individuals. Manners which had been formulated for the control of the court worked their way downwards through the middle class to become eventually the 'natural' behaviour of all classes. Eating with fingers, drinking soup without spoons or spitting on the floor became unacceptable for all social classes.

A parallel has been drawn between the development of diet as a secular science of consumption and of table manners as a process of civilization. The evolution of both entailed a regulation of desire, but their objects were rather different. Etiquette regulates the play of emotions within social groups by establishing norms of interpersonal behaviour. The traditional diet was directed at regulating physiological processes, which were seen to be the springs of irrational passions. In the disease metaphor of social pathology, unregulated appetites produced the unregulated society. There is, however, an important transition in the social significance of dieting from the eighteenth-century world of Cheyne and the world of mass consumption. In a consumer culture, the body assumes a new social and individual significance. It becomes

the site of personal strategies of health. Jogging, slimming and keep-fit programmes are designed to promote health as the basis of the good life. These
instrumental strategies of health are enthusiastically supported by the state as
the principal basis of preventive medicine. However, in the absence of a coherent system of communal, religious values, there is no obvious answer to the
question 'Health for what?' In the world of Cornaro, Cheyne and Wesley, the
individual exercised a stewardship over the body under the eye of God. The
religious calling in the world included a responsibility for personal health as
the basis for other achievements – the mastery of the soul and the passions. In
a society where such general religious notions play very little part in general
culture, health becomes itself the justification for dieting. The modern location of dieting is almost the reverse of its eighteenth-century position. The
purpose of the modern diet of consumer society is the production of desire –
the preservation of life to enhance the enjoyment of pleasures, the increase of
sexuality and the extension of enjoyments:

> Within a consumer culture the body is proclaimed as a vehicle of pleasure: it is
> desirable and desiring and the closer the actual body approximates to the idealised
> images of youth, health, fitness and beauty, the higher its exchange-value.
> Consumer culture permits the unashamed display of the human body.
> (Featherstone, 1982, pp. 21–2)

Whereas religio-medical dieting sought to achieve the control of the inner
body – the digestive roots of passion – by purgation and restraint, the consumer diet seeks to enhance the surface of the body – the cosmetic signs of
desirability – by the practices of body-maintenance.

Throughout this study, I have been drawing attention to the play of words
and the play on words: in particular, the political connotations of 'regimen',
'diet' and 'asceticism' have been considered at various stages. The ordering
of the surface of the body provides an equally interesting case of the political implication of the notion of 'cosmetics' and the growth of a new science,
cosmetology being 'that branch of applied science which deals with the external embellishments of the person through the use of cosmetic products and
treatments' (Wall, 1946, p. 28). 'Cosmetics' comes from '*kosmetikos*' or
skilled in adorning and ordering the body. Wall (1946) suggests the term
derives ultimately from 'kosmos' as the order, harmony and arrangement of
the body. Cosmetics was an important aspect of Greek culture and medicine
from an early time. Within the royal court, there was a group of specialists –
the *cosmetae* – specifically responsible for the adornments of royal women.
In addition, Galen's medical treatises included an outline of a medical science of cosmetics, based on an earlier work by Crito. From Greece, the
cosmetology of the body was transferred by Arab philosophers to Spain
and Europe, where cosmetics was finally separated from surgery in the fourteenth century by Henri de Mondeville and Guy de Chauliac. Under
Christianity, cosmetics meant vanity and adornments of the body were
regarded as temptations. John Wesley regarded the wearing of ornaments as
sufficient grounds for dismissing women from his chapels. In general, the
whole Reformation movement regarded cosmetics as an idolatry of the flesh,

but also associated such practices with idleness and an aristocratic way of life. While the Reformation brought the regimen of the internal body into the orbit of everyday asceticism, the ordering of the surface of the body was condemned as both irreligious and as a manifestation of aristocratic idleness. The association of leisure, idleness and aristocratic life-styles was brilliantly portrayed in Veblen's *The Theory of the Leisure Class* in 1899. Cosmetics along with corsets signified the absence of work and the ability to consume conspicuously. It was not until the nineteenth century that beauty culture became scientific and also became a standardized commodity for a mass market. A variety of changes took place, notably the development of massage on a scientific system in Peter Ling's *Manual of Swedish Movements* in 1813, the use of light and electricity for skin treatments, and the introduction of permanent waving. By the early part of the twentieth century, cosmetology had begun to be recognized in America as a distinctive area and 'cosmetic therapy' began to be covered by legislation in America from 1919. The regulation of the exterior of the body had assumed a scientific and rational status. Cosmetics became at the same time increasingly democratized and universal, moving from the leisure class downwards with the arrival of the mass market for standardized commodities. In Britain, rouge, lipstick and eye makeup ceased to be the stigmatic advertisement of the prostitute (Kern, 1975).

Cosmetics are a universal practice in human societies but their role in Western society is utterly different from their traditional use in pre-modern societies. Within a traditional social context, cosmetic decorations symbolized the incorporation of the individual within the social group and gave expression to common values and communal practices. Cosmetics communicate traditional symbols and signs; changes in personal cosmetics normally signify changes in social status, age of social membership. Female-body decorations signified a woman's changing personal status from puberty through marriage to widowhood. By contrast, Western cosmetics are largely determined by commercialized fashion and by individualized sexuality. Cosmetics have lost their rootedness in the sacred cosmos which connected cosmetic therapy with medicine. Cosmetics have been secularized along with the secularization of society. Cosmetic decoration does signify sexuality, but sexual decorations have been standardized by the calculative hedonism which is required by mass production (Brain, 1979). Cosmetic practices are indicative of a new presentational self in a society where the self is no longer lodged in formal roles but has to be validated through a competitive public space.

### The Critique of Foucault

The aim of Foucault's epistemology is to argue that objects of knowledge are not things-in-themselves but discursive objects which are the products of rules of discourse. Thus we have to understand 'diet' as the product of medico-religious discourses on the body and its functions; 'obesity' is not an empirical characteristic of unregulated bodies, but the effect of a language about bodies. Foucault's epistemology is a critique of empiricism and his position is derived extensively from contemporary discourse analysis (Chua, 1981). Although Foucault's approach provides a powerful perspective on medical discourse, it involves certain difficulties which it shares with structuralist analysis of discourse in general. Some forms of discourse analysis reduce the individual agent/speaker to the level of a socialized parrot, which must speak/perform in a determinate manner in accordance with the rules of language. There are alternatives to this form of structuralism which do not entirely obscure the role of human agency (Smart, 1982). Such an alternative was presented by V. Vološinov in *Marxism and the Philosophy of Language* (1973). Against Ferdinand de Saussure, Vološinov sought to retain some notion of the autonomy and voluntarism of the speaker who 'orientates to language not as a system of invariant rules with which he must comply but as a field of possibilities which he is to utilize in concrete utterances in particular social contexts' (Bennett, 1979, p. 76). Discourses are not linguistic machines which routinely and invariably produce the same effects, but possible modes of social construction the consequences of which contain a large element of contingency. Foucault's analysis of knowledge/power tends to assume without analysis that discourse has general social effects. He assumes, for example, that simply because there was in mediaeval times a confessional discourse, all confessions were of a similar form and had similar consequences for individuals regardless of their gender and class (Hepworth and Turner, 1982). Despite Foucault's interest in power, knowledge is too frequently extracted from its social context, because discourse is assumed to operate almost independently of the social groups which are its primary carriers. By contrast, it is impossible to understand a dietary discourse outside the social context in which it was formed and independently of the social groups which were simultaneously the targets and bearers of discourse. Another way of expressing this criticism is to argue that there is always resistance to discourse – an argument which Foucault recognizes but frequently neglects. The contemporary notion of beauty as slimness has become the object of feminist criticism and of women's groups which simply reject such norms of beauty as commercialism (Chernin, 1981; Orbach, 1978). Foucault's theory of knowledge has in many respects the weaknesses which characterize any dominant ideology thesis; it conflates the question of the logic of discourse with the issue of its social effects. We cannot assume that because, for example, modern advertising is based on a discourse of cosmetics that consumers invariably embrace its rules of production. Discourses are never uniform in their effects or unified in content. There are, in any case, a plurality of discourses with competing regimens of the body. Although Foucault has derived much from Nietzsche in terms of the argument that it is only by language that we can know anything and that different languages produce different knowledges, Foucault underplays two crucial features of Nietzsche's view of language. First, Nietzsche noted that the human animal is not only a sign-receiving animal but a sign-inventing animal too. Secondly, 'it was only as a social animal that man acquired self-consciousness' (Nietzsche, 1974, p. 299). Neither discourses nor disciplines are free-floating, autonomous practices, but deeply embedded features of social groups, economic classes and political organizations. Discourses, as Nietzsche noted, are themselves the social consequences of the endless struggle of social groups to realize their will and their potentialities.

# 8

# Government of the Body

## A Mode of Living

In Western thought, the human body is an ancient metaphor of political institutions, and was the dominant mode of theorizing political behaviour up to the seventeenth century, when the doctrine of individual property rights was fully articulated. The metaphor of the body was particularly important in the theory of kingship. The king had two bodies, a material body which was subject to corruption and decay, and a spiritual body which was symbolic of the life of the community (Kantorowicz, 1957). Given the centrality of the king to the political continuity of society, the death of the king's mortal body did not involve the demise of the authority of his sacred body. Hence it was necessary to construct an effigy of the king which sat upon the top of the coffin that bore his mortal remains. Political authority was thus preserved within the fictional person of the king's effigy (Prestwich, 1980). An attack on the person of the king was thus an attack on sovereignty as such and violent retribution had to be taken against regicides (Foucault, 1979). The body was, however, used as a more general metaphor for the structure and function of society as a whole. The teleological purposiveness of the body was employed to legitimate political and social divisions in society: 'Society, like the human body, is an organism composed of different members. Each member has its own function, prayer, or defence, or merchandise, or tilling the soil. Each must receive the means suited to its station, and must claim no more' (Tawney, 1938, pp. 35–6). The parallel between Christ's charismatic authority as head of the church and the king's institutionalized authority was obvious enough. The church was a body of believers, welded together by the grace of Christ and ruled by episcopal powers. Like the king's body, the church was both a secular institution and a mystical body founded on the rock of faith. The body metaphor was thus a well-established feature of mediaeval thought and, although it was eventually replaced by the liberal discourse of individual rights, it survived in the so-called 'organic analogy' of early structural functionalism in the writing of Herbert Spencer, and it also survived in such legal notions as 'the corporation'.

The body/politics metaphor has enjoyed a widespread currency in such expressions as 'the body politic', 'the social body', 'the head of state' and 'the body of the church'. In this chapter, I want to extend this metaphor to consider the notions of 'the government of the body' and 'the anarchy of the body'. Having discussed the application of political discourse to the relationship between will and desire, I provide a theory of these controlling and anarchic functions by an exploration of the relationship between 'female diseases' and patriarchy.

In earlier publications (Turner, 1982a; 1982b), I have explored the concept of medical regimens as governments of the body with special reference to dietetics. In Greek medicine, diet (*diaita* or 'mode of living') referred to the general conduct and organization of life, including forms of dress, behaviour and attitudes. In its more restricted sense of a mode of eating, diet was an essential element of the Greek medical *regimen*. A medical regimen is a set of rules or guidelines imposed upon a client to secure his or her well-being. When the body is conceived as an input–output system, the regimen restores the equilibrium of the body through a regimen of purges, fasting, sweating and diet. Regimen also, of course, has the somewhat antiquated meaning of 'government' and is the root of 'regime', and 'regiment'. The *diaita* was a mode of living set within a particular government of the body by medical practices. We can envisage such regimens occurring along a voluntary/involuntary continuum. Voluntary governments involved a social contract between patient and doctor, whereby, in exchange for the medical fee, the patient contracted into a mode of living to achieve the restoration of health. Like other political contracts, the medical regimen involved a certain loss of self-will: the regimen works if it is followed. Involuntary regimens may be illustrated by enforced incarceration of the insane or the seclusion of lepers. Regimens imply therefore an element of choice and responsibility on the part of patients, but if we take a wider view of the whole process of nourishment of the body we need a more complex model.

Eating can be conceived as a fundamental 'body technique' (Mauss, 1979), that is, an activity which has a basic physiological function, but which is heavily mediated by culture. While feeding a child is an act of care and support, creating a bond between parents and child, it is also the imposition of a 'mode of living' (a regimen) on a subordinate. Gaining control over our own feeding patterns involves a growth in personal autonomy, and refusing to eat or engaging in forced vomiting is an act of rebellion. Although this regulation in symbolic terms is evidence of the operation of fundamental categories of thought (Lévi-Strauss, 1970), the mode of eating can also be seen as a site of familial politics. Thus there is the self-imposed diet as an illustration of a voluntary regimen and the body techniques which a child learns as the imposition of power across generations.

One might also think of the regimen imposed by nature on culture. Maturation and ageing are processes over which we have little direct control and which impose on us new modes of living. In old age and sickness, we 'naturally' require less to eat. These physiological processes are backed up by cultural expectations as to what is appropriate for different age groups or sexes. These contrasts suggest a property-space representation of the mode of living which may be voluntary/involuntary or externally/internally negotiated (see Figure 2). 'Physiological routines' refer to 'natural' patterns of hunger and thirst which are imposed on us by our membership in nature. Clearly these routines are themselves partly negotiated, but there are limits beyond which we cannot stray. An internal voluntary mode of living would be simply a decision to reduce our calorie intake as a personal choice. Refusal to eat may also be an act of rebellion (as in the case of IRA suspects during Internment). Forced feeding is the opposite, namely an act of external terror being an invasion of the body. Medical regimens are voluntary but also external social contracts between parties as an exchange. In these examples, I am arguing that the body is a location for the exercise of will over desire. The achievement of personal control over diet is an act of will which enhances self-esteem, but it can also be imposed from without as a denial of will. The achievement or exclusion of certain modes of living can be suitably analysed by political metaphors of government – from dictatorship to anarchy.

|  | Involuntary | Voluntary |
|---|---|---|
| Internal | Physiological routines | Diet |
| External | Forced feeding | Medical regimens |

Figure 2

## The Orgy and the Fast

From a rational standpoint, the body has been traditionally conceived as the source of irrationality, as a threat to personal stability and social order. The sexuality of the body in particular is a threat to orderly succession and family authority. As we have seen in earlier chapters, much of Western cultural history has been seen as a pendulum swing between orgy and asceticism, the forces of Dionysus and Apollo. As Borges (1973, p. 83) has observed.

> The world we live in is a mistake, a clumsy parody. Mirrors and fatherhood, because they multiply and confirm the parody, are abominations. Revulsion is the cardinal virtue. Two ways (whose choice the Prophet left free) may lead us there: abstinence or the orgy, excess of the flesh or its denial.

The orgy and the fast in many ways summarize much of the cultural history of Western Christian civilization. Orgiasticism has often been associated with political protest, while asceticism has been connected with restraint and control. Gouldner (1967) associated the Dionysian cults with the marginal, disprivileged groups in society (peasants, women and slaves). Most of these commentaries associated orgy and fast with the macro-politics of the social system. For example, Bakhtin (1968) connects peasant revolts with the role of festival as an oppositional force to the norms of individualism which were set within the market in the work of Rabelais. However, in the modern world of consumerism, we can also think of two medical conditions – bulimia and anorexia nervosa – as two individualized forms of protest which employ the body as a medium of protest against the consumer-self (see Figure 3). Orgy and ascetism are both culturally mediated 'modes of living' with a specific social significance. Orgiastic release is typically a protest against political controls. As we shall see in later chapters, the orgiastic festival of mediaeval Europe is clearly separated from modern calculating hedonism.

If orgy and asceticism are culturally mediated and social activities, bulimia and anorexia nervosa are individual solutions to social problems and they are more closely dominated by the routines of physiology. While both bulimia and anorexia are individually selected solutions to familial crises, they have unintended physiological consequences over which, by definition, the individual has no control.

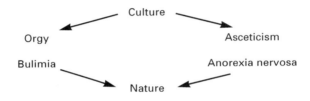

Figure 3

**On Disease**

The relationship between 'disease' and 'illness' raises a variety of basic questions about the cultural and social mediation of human physiology. From a logical point of view (F.K. Taylor, 1979), the term 'disease' has a specific technical sense referring to 'configurations of pathological abnormalities'. By contrast, 'illness' refers to clinical manifestations which can be regarded as either symptoms (subjective sensations) or signs (objective findings discovered by an expert observer) (Feinstein, 1967). It follows that illness has an irreducible social component involving subjective responses of patients and diagnostic judgements of professionals. Both of these are clearly subject to social determinations (Bloor, 1976). The social constitution of illness as a classification has been expressed strongly in the case of mental illness by Morgan (1975) who argued that, while 'disease' refers to all living species, 'illness' exists only in society. This notion of the essentially social character of illness can be further extended by reference to the 'sick role' (Parsons, 1951) or 'illness behaviour' (Mechanic and Volkart, 1961). Illness can be regarded as deviant behaviour, but it is heavily structured by cultural categories which legitimate or normalize deviance as a medical condition. Clearly not all persons experiencing illness symptoms seek out professional medical help. Once a sick role is adopted the management of illness is subsequently held to be the mutual responsibility of both patient and physician. These mutual responsibilities form the basic element of a medical system.

What is missing in medical sociology is the recognition that the search for the social and/or physiological aetiology of disease and illness locks into a deeper and more persistent set of theoretical issues which links together philosophy and sociology. The concept of illness in particular brings together three fundamental debates which have shaped sociology from its inception, namely the relationships between nature and culture, individual and society, and mind and body. It is sometimes implied that only a peculiar class of illnesses actually raise these issues, namely those which are referred to as 'psychosomatic illnesses'. From a sociological point of view, it is difficult to accept even this notion, since to have any illness is the effect of diagnostic processes and professional judgements, which are in turn the outcome of historical and social determinations. From a Foucauldian perspective, the labels of scientific medical practice are not statements about 'real' disease entities, but effects of power-knowledge and products of specific discourses (Foucault, 1967). Thus, even death, that final arbiter of physiology, is a social category and not an innocent medical tag (Ariès, 1974). Although the truth or falsity of propositions may be independent of their social context, there are no beliefs which do not have social causes. Medical beliefs are thus proper objects of a sociology of knowledge.

**Man is What He Eats**

It has been suggested that enquiries within medical sociology often mask deeper issues within traditional philosophical debates. One exception to this claim is *Awakenings* in which Sacks (1976) explores the sequelae of Parkinsonism via Leibniz's theodicy. In this chapter, I argue that anorexia nervosa perfectly illustrates the underlying philosophical bases of sociology (individual/society, nature/culture, mind/body). To develop this illustration, anorexia is located within a theme which emerged in early Marxism through the medium of Feuerbach's sensualist epistemology under the slogan 'Man is what he eats' (Cherno, 1962–3). While these philosophical issues are endemic to medical sociology, anorexia is peculiarly suited to my argument.

There is some agreement (Palmer, 1980; Kalucy, Crisp and Harding, 1977) that the first clinical descriptions of anorexia nervosa appeared in France and England in the late 1860s. In England, Sir William Gull delivered an address at Oxford in 1868 on the characteristics of anorexia, which he elaborated in 1873 by describing the case of Miss A. (Gull, 1874). In the late 1880s, the work of Charcot in Paris suggested that anorexia was one feature of the hysterical syndrome (Charcot, 1889), and this theme was further elaborated in the work of Freud and Breuer in the 1890s in their analyses of

Frau Emmy von N. and Fraulein Anna O. (Freud and Breuer, 1974). The historical specificity of the eruption of anorexia in the late nineteenth century suggests a connection with Foucault's idea of the 'hysterization' of women's bodies (Foucault, 1981) and the peculiar conjunction of social structures which produced a crisis in middle-class, urban family life (Janik and Toulmin, 1973). This conjunction of circumstances was combined with a specific interest in family organization by the medical profession (Donzelot, 1979).

Like hysteria, anorexia is clinically almost entirely an illness specific to women. It is estimated that only one in ten 'victims' of anorexia nervosa is male (Palmer, 1980). The gender specificity of the illness is also suggested by the temporal nature of its onset, namely in the period between puberty and menopause. More precisely, anorexia typically develops at the age of fifteen and most cases are identified before the age of twenty-five. In short, the illness is characteristic of young women. There is also evidence that 'anorexia nervosa' is becoming an increasingly popular diagnostic label amongst medical professionals which is indicated in the increased prevalence of the illness (Crisp, Palmer and Kalucy, 1976). Much of the literature on the illness has been motivated by a feminist critique of the position of women in society in relation to the sexual division of labour and in relation to the patriarchal system of professional medicine. In this view, women are peculiarly subject to the contradictory expectations of beauty in a consumer society where male criteria of aesthetics predominate (Chernin, 1981). Anorexia raises the question of whether the human body – its size, weight, gestures and deportment – is shaped in accordance with cultural criteria of appropriateness. In this regard, Mauss's notion of 'body techniques' as 'the ways in which from society to society men know how to use their bodies' (Mauss, 1979, p. 97) is of major importance.

Medical definitions of anorexia nervosa indicate the uncertain and complex nature of the illness. Anorexia appears to hover insecurely within a nosological map; it is partly a disease and partly an illness. Feighner (1972) identified a cluster of clinical signs and symptoms as necessary for the diagnosis of anorexia nervosa: (1) age of onset prior to twenty-five years; (2) at least twenty-five per cent loss of original body weight; (3) a distorted attitude towards food and eating; (4) no known prior medical illness which could account for anorexia; (5) no other known primary affective psychiatric disorders; (6) at least two of the following – amenorrhoea, lanugo, bradycardia, overactivity, bulimia and vomiting. Because weight loss and menstruation problems may be absent and sporadic in anorexic patients, some authors have suggested 'dietary chaos syndrome' as a more appropriate diagnostic label (Palmer, 1979). Alternatively, Bruch (1978), Crisp and Toms (1972) and Russell (1970) have adopted entirely psychological criteria for diagnosis, namely the relentless pursuit of thinness associated with a fear of fatness. From a sociological point of view, what appears significant about anorexia is that it is impossible to detach it from a social aetiology, criteria of social deviance and social symbolism. Some of the more promising interpretations of anorexia perceive it in terms of a struggle within the middle-class family where, over-protected daughters seek greater control over their bodies and therefore their lives (Bruch, 1978). There is a sense, therefore, in which girls 'choose' anorexia as a deliberate personal strategy of autonomy. The so-called 'control paradox' of anorexia can often be seen in near-religious terms as asceticism, that is 'an attempt to attain spirituality or goodness' (Lawrence, 1979, p. 96) through the subordination of the flesh. Yet the deeper paradox of anorexia is that this attempt to control the body results in its dominance – food, eating, vomiting, slimming become all-consuming passions. Within a broader cultural perspective, the asceticism of Buddhist practice selected a middle way between extreme asceticism and hedonism. Mogul (1980) comparing the asceticism of Gautama Buddha and the modern anorexic noted that mortification of the body leads, not to personal freedom from its needs, but to mental enslavement.

The sexual symbolism of anorexia is equally contradictory. Autobiographical accounts by anorexic women indicate that extreme slimming is associated in puberty with a rejection of sexuality through the suppression of menstruation. Obedience to parental controls with regard to moral purity may also be involved in this rejection of boyfriends, sexual maturity and sexual experience. At the same time, our current cultural norms of female beauty emphasize slimness and slenderness in contrast to the weighty matrons of Rubens and Rembrandt. While denying the physiology of her sexuality, the anorexic woman conforms to the accepted standards of female attractiveness (Orbach, 1978). Whereas Christian asceticism aimed to liberate the soul by subordinating the flesh, the body maintenance theme of modern consumerism aims at enhancing pleasures. The thin body is better equipped for desire. By denying her sexuality as a personal choice, the anorexic accepts, or at least conforms to, an ethic of consumer sexuality. It is interesting to note, therefore, that ballerinas who are an epitome of sexual attractiveness are, as a subculture, also commonly anorexic (Druss and Silverman, 1979).

The paradoxes of anorexia reproduce, to use a phrase from Lukács (1971), the antinomies of bourgeois thought. As I shall argue more fully shortly, it is a search for individual freedom and individuation from the 'golden cage' of the middle-class, over-protective family and a quest through the rigours of secular asceticism for personal perfectability. At the same time, it is over-determined by the culture of narcissism, consumerism and the patristic norms of slender femininity. It is an act of adolescent rebellion against parental control, but it ends in physical enslavement to the body – a rebellion that often ends in death. This individual act of self-assertion ironically reproduces the conventional social criteria of youthful female beauty. Anorexia can be seen as an exercise of mind over body, of culture over nature. Yet the loss of weight, deformities of bone structure, loss of menstruation, hyperactivity, malnutrition, hypersthenia and anaemia are consequences which cannot be readily controlled. This is the reassertion of natural over cultural processes. Anorexia can thus be meaningfully described as a 'psychosemantic fallacy'; it is

> a disease in which the concept of the whole person is so confused, so dialectically divided, that 'I' can at the same time be choosing to live, as the self, and choosing to die, as the body, however unconscious those choices may be . . . both suicide of the schizoid type and anorexia nervosa involve a denial of reality which depends upon an acceptance of a split between self and body, and is only possible through paradox. (MacLeod, 1981, p. 88)

What I want to suggest in this chapter is that, to achieve a wider and a firmer theoretical grasp of these paradoxes, it may be valuable to reconsider some issues relating to the nature–culture contrast in the debate between Marxists and Feuerbachians.

## Sensualism

Feuerbach's critical criticism is now largely remembered as simply the point of departure for Marx's early attempt to develop a theory of *praxis*. Both Marx and Engels embraced Feuerbach's critique of religion as the starting point of all criticism because it represented an advance on French materialism which they regarded as mechanical and static. Marx came to reject Feuerbach because Feuerbachian criticism remained purely cognitive and contemplative, that is, the overcoming of human alienation was to be achieved merely by passive cognition not by practice. Thus in the first of the 'Theses on Feuerbach' we read that

> Feuerbach wants sense objects really distinct from the objects of thought, but he does not conceive human activity itself as *objective* (*gegenstandliche*) activity. Hence, in the Essence of Christianity, he regards the theoretical attitude as the only genuinely human attitude, while practice is conceived and fixed only in its dirty Jewish manifestation. Hence he does not grasp the significance of 'revolutionary' or 'practical-critical' activity. (Marx, 1976, pp. 61–2)

Feuerbach was criticized by Engels for replacing the revolutionary emancipation of the proletariat with the liberation of people through love. Thus, for Feuerbach, 'sexual love finally becomes one of the highest forms, if not the highest form, of the practice of his new religion' (Engels, 1976, p. 29). Feuerbach's sensualism was thus dismissed as cognitive, subjective and individualist. For Marx, the subject of the interchange between man and nature is the social collectivity which has a history and a specific form; it is not the individual sensuous being but the structured social collectivity within which individuals labour and reproduce (Hanfi, 1972).

As many commentators have since observed, the break between Marxist materialism and Feuerbachian anthropology was never as clean and neat as Engels wanted to suggest; furthermore, the break between Marx and Feuerbach diminished Marxism by eliminating any conception of people as sensual, emotional beings, as entities which paradoxically have bodies and are bodies. Marx's concept of *praxis* and dialectic as the solution to the mind/body dichotomy, as expressed in the sterile opposition between mechanical materialism and active idealism, grew out of Feuerbach's project to liquidate the rationalist prejudice of Cartesianism which suppressed the emotional and passionate dimension of human existence. There is something odd in the view that Feuerbach's solution to the alienation of human essence in abstract theology was merely cognitive, given Feuerbach's view that thinking and experiencing are united in sensuous practices. In part, this rejection of Feuerbach can be associated with an implicit asceticism, particularly in Engels, which rejected any argument in favour of the centrality of desire in human relationships as an example of the decadent utopianism of Charles Fourier (Beecher and Bienvenu, 1972). The idea that any revolutionary reconstruction of society would also have to entail a fulfilment or enhancement of human sensual satisfaction, particularly sexual enjoyment, largely disappeared from later Marxism: revolutionary asceticism became opposed to bourgeois corpulence (Schmidt, 1971). The body, to employ an Althusserian metaphor, ceased to be an object of Marxism theoretical labour. The principal exception to this observation can be found in the work of Sebastiano Timpanaro (1970) for whom the frailty of human existence represents in death the final triumph of nature over history. The attempt to retrieve the passionate life of the body in modern social theory has occurred through neo-Freudianism in critical theory (Marcuse, 1969) and through Nietzsche in structuralism (Benoist, 1978).

My argument is that neither Marxism nor sociology has in recent times attempted to produce a theory of the body; it is unfortunate that this absence in social theory has been seized upon by the proponents of sociobiology which is largely reductionist in approach. To some extent, this is an oddity since both Marxism and sociology can be said, at a metatheoretical level, to have been constructed on the attempt to resolve the philosophical problem of mind/body. We have seen that the young Marx set out, via Feuerbach, to transcend mechanistic materialism which ultimately reduced conscious activity to physiology. Equally sociology can be seen as an attempt to transcend Kantian epistemology in which man is both an entity within nature (the phenomenal world) and an active member of a moral community (the noumenal world). Durkheim's analysis of totemism and Weber's interpretative sociology were both grounded in this Kantian problem. In the long run, any attempt to incorporate a theory of the body in mainstream sociology has, however, been regarded as the thin end of biological reductionism (Parsons, 1937).

In this chapter, I am mainly concerned with a possible Marxist solution to this absence through Feuerbach rather than with sociology. One Marxist solution could be located in the notions of 'practice' and 'contradiction'. Specifically my argument is that the body is both a means of labour and an object of labour; we realize ourselves through labour on our bodies and this labour on the body is a social practice. From this notion of body practices, my aim is to work towards the claim that illness is not simply an event that happens to the body, but paradoxically a choice. Before coming to this argument, it is important to examine further aspects of Feuerbach, particularly his later views on nutrition as the secret of the nature/culture separation.

In his later works, that is from 1848 to 1862, Feuerbach was preoccupied with the relationship between nature and human existence, and in particular with the paradox that human beings are both in and of nature (Wartovsky, 1977). By redirecting the arguments of *The Essence of Christianity* through physiology, Feuerbach attempted to overcome the object/subject dichotomy by arguing that external nature becomes internal subjectivity through the incorporation of nature through eating. External reality is not given but acquired through sensuous practical activity. We are linked to external reality by our physiological needs, but this is an active linkage since external reality is literally appropriated and internalized by consumption. This dependence on nature was mythologized in religion as a dependence on the Father, but in his later work, under the influence of Jakob Moleschott's *Lehre der Nahrungsmittel* (Theory of Nutrition) of 1850, Feuerbach's original sensualism was converted into a specifically physiological doctrine. Men are the products of what they eat, so that Feuerbach's notion of sensuous practice was lost; human beings become the passive consequences of organic processes. These arguments were again employed in the analysis of religion where, in *The Mystery of Sacrifice or Man is What He Eats* of 1862, Feuerbach noted the incorporation theme in all communal religious sacrifices. The origin of sacrifice lies in human dependence on external reality, but the result of sacrifice is a false sense of confidence and independence. It is easy to detect massive problems in Feuerbach's materialism and equally easy to illustrate its absurdity as when, quoting Moleschott, Feuerbach saw the consumption of beans as the solution to revolutionary transformations in Europe (Kamenka, 1970, p. 112). However, as most interpretations of Feuerbach indicate, it is possible to extract valid principles from his emphasis on sensuous activity. What Feuerbach never entirely lost sight of was the premise of *The Essence of Christianity* that man is not simply a cognitive being, but a sensuous, active agent both in relation to the external environment of nature and to the internal environment of his sensations and sensibility. Wartovsky (1977, p. 408) observes this premise especially clearly in Feuerbach's attempt to overcome the mind–body identity:

> The identity is not, reductively, that of the mind with the body, *as body*; nor of the body with the mind, *as mind*; rather the identity, or the unity, is the *totality itself*; as a functional or organic one, that is as an *activity* of living, thinking, feeling, willing, whose organic condition is certainly a material or physical body, but only a body of a certain kind, the *acting* body, *whose externality is a relational one*, and that therefore cannot be reduced to a composite or aggregate physical thing except in death.

We can thus stretch Feuerbach's rendering of Moleschott as being the idea that 'man is himself a product of his productive activity, that he creates himself in the labour of producing the means of his subsistence and in the social organization that assures this production and reproduction of species existence' (p. 413).

What is generally absent in Feuerbach's sensualist materialism is any developed sense of the social conditions which are combined with, and a condition of, physiological reproduction. Feuerbach takes a one-sided view of eating as an individual incorporation of nature but,

> In a fully two-sided dialectic, not only the eating of food, but the obtaining, the production and reproduction of the means of existence would have to be taken into consideration. The dialogue between my stomach and the world, in real activity, is mediated by the dialogue between production and consumption, the social dialogue of human praxis that Marx developed in his political economy. (p. 416)

It is in this respect interesting, for example, to compare Feuerbach's view of sacrifice as alienated consciousness of dependency with those put forward by William Robertson Smith, Emile Durkheim and Marcel Mauss (Hubert and Mauss, 1964). For later sociologists, the explanation of sacrifice was to be found in its social not personal functions. Thus, the ritual sacrifice and consumption of the god created a bond between people and god, while at the same time producing a social bond within the group. Although Feuerbach's sensualist philosophy suffers from a variety of defects and failures, it is possible to extract an important core of ideas from his attempt to transcend certain traditional problems in philosophy.

**Body Practices**

In the following section I want to outline some general principles for the analysis of anorexia nervosa which follow from Feuerbach or are compatible with revisions of Feuerbach's approach. First, bodies are objects over which we labour – eating, sleeping, cleaning, dieting, exercising. These labours can be called body practices and they are both individual and collective. These practices tie us into the natural world, since our bodies are environments, while also locating us into a dense system of social norms and regulations. Much of the work of Erving Goffman, for example on face-work (1969), can be seen in this context. The body is a site of enormous symbolic work and symbolic production. Its deformities are stigmatic and stigmatizing, while at the same time its perfections, culturally defined, are objects of praise and admiration. Because of its symbolic potential, the body is also an object of regulation and control through asceticism, training or denial as I have argued in 'The government of the body' (Turner, 1982a). It is possible to trace this regulation of the body in public space as a civilizing process, as in Norbert Elias's analysis of etiquette (1978), or as a process of rationalization as in Michel Foucault's analysis of discipline (1979). However, following Feuerbach, I want to argue that the body is both an environment we practise on and also practise with. We labour on, in and with bodies.

Secondly, to emphasize an argument which is lacking in Feuerbach, our body maintenance creates social bonds, expresses social relations and reaffirms or denies them. While Robertson Smith's analysis of religious groups has often been criticized, he can be said to have expressed an essential sociological insight into social bonding. For Smith, to understand religion we have to understand religious practice and to understand religious practice we have to consider its social effects; in particular, religious rituals create a social bond between their practitioners. There is an essential relationship between the sociology of religion and sociology as such. Religion or *religio* means to bind together. Sociology from *socius* is the study of fellowship or more generally the processes that cement and corrode social relationships. Following this etymological line, the elementary forms of social relationships in society are exchange relationships between parents and offspring which through socialization create bonds of dependency. The most basic of these dependency relations is located in the exchange of food from parent to child that creates binding and obligation. To grow up is to achieve individualization through self-autonomy, especially control over personal body maintenance. Paternalistic authority grows out of and is legitimated by these dependency relations, which emerge directly out of these body-maintaining services rendered by parents to their offspring. Refusal to eat, vomiting and dietary disorders in this respect are precisely disorders. In the absence of language, vomiting on the living-room carpet is one clear statement of resistance to parental control. Similarly, refusing to eat is an opposition to parental feeding which gives the child some control over body functions.

Thirdly, as Wartovsky notes in his criticisms of the ahistorical nature of Feuerbach's attempt to translate sensualism into dietary materialism, this micro-politics of the body has to be set within the wider context of production and reproduction. I have suggested that Marxism has largely neglected the reproduction of bodies and population in favour of a primary concern for the production of commodities. It can be argued that Foucault has taken the problem of the reproduction of the body of individuals and the body of populations as a central focus of his philosophy of power/knowledge. More generally, structuralism has taken the conflict between power and desire as the crux of all forms of authority; the conflict between the id and the superego is thus a model of political struggles. There does appear to be at least a vague resemblance between Feuerbach's attack on Cartesian rationalism by asserting the primacy of sensuous need over theology and Foucault's attack on rationalist knowledge by asserting the autonomy of desire. To render these debates into a sociology of the body requires the recognition that it is the female body which, historically, is the focus of social control through knowledge and authority, because women produce men, while men control women and property as commodities. It can be argued that men control the distribution of property (under primogeniture for example) by controlling women ideologically (such as the pre-modern doctrine of women as a deformity or secondary creation) through the institutional apparatus of the family and the state. Authority is thus legitimated through the denial of desire under a variety of paternalistic institutions (Sennett, 1980). My argument is that we should not see anorexia nervosa as an isolated epidemic of modern society, but rather treat it as part of a complex collection of female disorders in the context of changes within Western society over at least the last hundred years.

## Contradictions

The intention here is to take these three dimensions – choice, dependency and social context – to illustrate a series of contradictions within anorexia nervosa. Most of these illustrations are taken from the autobiographies of anorexics, backed up by qualitative clinical data. Various authors have concentrated on family background in the aetiology of anorexia. Two salient features have emerged from these studies, namely (1) an overpowering, dominant mother involved in an excessively controlling relationship over the daughter where there is a contradictory emphasis on compliance, cleanliness and competition, and (2) inadequate preparation for adolescence, because there are few opportunities for the individualization of the child, particularly for sexual and gender identity. In more detail, anorexic families are close knit, small (with an average of 2.8 children), and achievement orientated. There is a paucity of sons in such families, that is an estimated sixty-six per cent have daughters only. The mothers of anorexics tend to be themselves preoccupied with weight problems, their own careers and in general raise their children to satisfy their own interests (Bruch, 1978). There is a contradictory relationship between the emphasis on success and competition which requires some independence on the part of the child in the outer world of school and university, and the emphasis on obedience and compliance which subordinates the individuality of the child to that of the mother. Within the middle-class context of anorexia, the child is faced with the possibility of failure at school, inability to match parental expectations and the symbolic disobedience of educational failure. It is interesting therefore that the onset of anorexia often corresponds to entry into higher education. The title of Bruch's classic study of anorexia – *The Golden Cage* – perfectly captures the feeling of the anorexic daughter: 'She was like a sparrow in a golden cage, too plain and simple for the luxuries of her home, but also deprived of the freedom of doing what she truly wanted to do.' Self-chosen starvation provides the hope of escape from the cage, but it is a paradoxical bid for freedom.

By suppressing menstruation, the daughter suppresses sexuality and adopts a permanently childlike body and attitude to the mother. At one level, anorexia is a refusal to mature. At the same time, self-starvation gives an enormous sense of self-control via control of biological processes. Food refusal 'is a defence against the original fear of eating too much, of not having control, of giving in to their biological urges. . . . This accumulation of power was giving her another kind of "weight"' (pp. 4–5). Sheila MacLeod's autobiography *The Art of Starvation* (1981) documents how the 'choice' of anorexia often brings with it an early feeling of elation and independence as the anorexic experiences the personal pleasure of control through personal discipline. Anorexia is chosen as a defence against confusion between opposites – compliance/independence, maturity/ childhood, sexuality/neutrality. While the anorexic cannot adequately control this exterior world of contradictions, she can at least control herself through the ascetic regime of anorexia; this is her peculiar, compelling path to self-hood. However, once anorexia has been chosen as a committed identity, there are aspects of bodily processes which then assert their own logic and autonomy on the newly independent self. It becomes increasingly difficult to control, interrupt or redirect the process of weight loss, absence of appetite, overactivity, insomnia and amenorrhoea. There may also be extensive dental decay and loss of teeth. The anorexic may also become locked into the contradictory cycle of starvation, bulimia, guilt and vomiting. In many respects this pattern of contradictory behaviour is reminiscent of a theological, moral disorder known in Catholic pastoral theology as scrupulosity. This refers to the obsessive behaviour of children who attempt to adhere to every detail of ritual and moral codes presented to them by authority figures (Hepworth and Turner, 1982). Because they set themselves impossible tasks of rigid conformity to rules, they must necessarily fall into sins which produce guilt and the desire to impose even stricter forms of conformity. They are involved in a moral spiral from which there is little escape. Similarly, the anorexic pattern of asceticism requires obligations which cannot be met so that lapses into self-indulgence are regarded as imperfections which drive the sufferer into further enforcements of the regimen. Thus, an initial act of governing the body to achieve identity and autonomy is replaced by an anarchy of the body which denies the will of the subject/victim, whose response is an intensified programme of dieting and exercise.

   A similar account of this process is to be found in Aimée Liu's *Solitaire* (1979). This autobiography also reflects the paradox of a government and anarchy of the body. Like MacLeod, she was a child acutely aware of her sense of powerlessness within a middle-class family which was socially successful. Dieting was her 'first totally independent exhibit of power' (Liu, 1979, p. 36). While a successful girl at school, Liu was overwhelmed by her personal sense of failure and inadequacy. Despite her attractive looks, she was horrified by sexuality, partly as a result of sexual harassment as a child. Her control of menstruation was thus a triumph: 'My periods have stopped! I don't suppose the reprieve will last forever, but for the moment it delights me. And the more weight I lose, the flatter I become. It's wonderful, like crawling back into the body of a child' (p. 41). Diet, then, is one of the few areas of personal control and discipline which young women from close, but competitive family environments can exercise as an act of personal autonomy: 'My diet is the one sector of my life over which I and I alone wield total control' (pp. 46–7). These two autobiographical accounts of anorexia are particularly interesting in terms of the dominance of a political discourse – control, rebellion, discipline, autonomy, choice – by which they attempt to render the physical experience of dieting.

Much of the contemporary literature on anorexia argues that it is at least in part caused by the prevalence of cultural norms advocating thinness as a personal value and that these norms in turn reflect the dominance of patriarchal values and patriarchal authority over women (Chernin, 1981; Lawrence, 1979; Orbach, 1978). While there is much to be said in favour of such an interpretation, the onset of anorexia is situated in a conflict over dependence and autonomy in the relationship between mother and daughter. In this context, the refusal to eat, however secretly that refusal is pursued, is an act of rebellion which breaks the social bonds created by nurturing. Some of the earliest accounts of the illness in Freud and Breuer's studies of hysteria drew attention to this aspect of rebellion. In the case of Frau Emmy von N., Freud's attempt to encourage her to eat reproduced the rebellion she had experienced as a child: 'the furious look she cast me convinced me that she was in open rebellion and that the situation was grave' (Freud and Breuer, 1974, p. 141). Frau Emmy reported that, 'when I was a child, it often happened that out of naughtiness I refused to eat my meat at dinner. My mother was very severe about this and under threat of condign punishment I was obliged two hours later to eat the meat, which had been left standing on the same plate' (p. 141). In *The Art of Starvation*, MacLeod also emphasizes that the refusal to eat is especially potent as a rejection of the mother as the source of food and life: 'What better revenge can there be on an unfaithful mother who gives her body to another than to reject her, and with her the principle of nourishment, in becoming anorexic?' (MacLeod, 1981, p. 35). Anorexia thus transforms the previously compliant 'good girl' into a naughty but determined rebel. The rebellion is of course primarily a symbolic gesture which cuts off the nurturing bond which the girl experiences as bondage. This aspect of naughtiness is frequently commented on in the literature, but the significance of it is equally frequently lost.

Anorexia involves a power struggle within the family over food, with the parents attempting to force their daughter to eat. It is obviously difficult to force a child to eat, and in any case, the disobedient daughter can resort to secret or deviant tactics – vomiting and punitive exercises. As Bruch (1978, p. 2) notes with respect to one of her anorexic patients, 'Formerly sweet, obedient and considerate, she became more and more demanding, obstinate, irritable and arrogant. There was constant arguing not only about what she should eat but about all other activities as well.'

The girl's search for individuation and autonomy is thus fought out in a political language of opposition to the bonding created between family members by the common table. The political metaphors of medical language are particularly interesting in this respect. A regimen is a government of the body and the forms of eating imposed by parents on their children can thus be seen as an aspect of domestic government or a regime for the control of bodies. Anorexia is not simply a disorder of metabolism, but a dis-order of social relations. Anorexia is an alternative, disruptive regimen, an anarchy within the domestic government. But, as I have already suggested, anorexia becomes an anarchy of the organic system which imposes its own logic and autonomy. The search for autonomy becomes an illness which imposes its political authority over the body of the victim.

It is interesting to compare anorexia nervosa with Durkheim's account of egoistic suicide. For Durkheim (1970) egoistic suicide appears to be sub-jectively the choice of an individual to terminate their life, but sociologically it is the product of the weakening or collapse of social bonds linking the individual to the social group and thereby exposing them to the destructive suicidal forces of an individualistic culture. By contrast, the suicide of anorexics appears to be, in part, the product of over-socializa-tion, of social bonds within the family that create over-protective surveillance and discipline (Mannoni, 1973). The suicidal path of the anorexic can thus be interpreted as the result of too much rather than too little parenting, but since this over-socialization is brought about by the mother in a domestic government where the father is typically absent or weak, the anorexic household suffers from matriarchal not patriarchal con-trol. This kind of argument appears therefore to depart from one interpretation of anorexia which regards the illness as a consequence of patriarchal values enforcing unworkable norms of slender beauty on women. The problem of female forms is further reinforced by a commercial capitalist system which promotes commodities by reference to a body aes-thetic. The value of this feminist critique is that it locates anorexia within the context of the general position of women in society as an historical issue. In this respect, anorexia has to be seen along with hysteria in the nineteenth century and depression in the twentieth as an illness which gives expression to the structural limitations placed on women who are at the same time, especially in the middle class, expected to be successful in the public domain. To understand anorexia therefore, it will not be sufficient to situate it simply within the space of the private household, since these pri-vate spaces are themselves determined by the wider structure of industrial society.

To return to Feuerbach, if we take him literally – Man is what he eats – we may wish to argue that Woman is what she looks like. A woman's form is sym-bolic of character. The obese woman is not simply fat; she is also out of control. The unrestrained body is a statement or a language about unre-strained morality. To control women's bodies is to control their personalities, and represents an act of authority over the body in the interests of public order organized around male values of what is rational. There is a good argu-ment for examining anorexia alongside a more general notion of the restraint of women's bodies. While it may appear to be a bizarre comparison, I want to conclude this discussion of the government of the body by an examination of the history of corsets in the nineteenth century.

**Corsets**

Tightlacing to achieve a constricted waistline dominated British fashion from the 1830s to the 1890s. The unrestricted body came to be regarded in this period as symbolic of moral licence; the loose body reflected loose morals. At the same time, the corset was an emblem of a leisure class since a corsetted woman was unable to perform manual labour. A number of pressures – moral, economic, status, fashion – forced or encouraged women to shape their bodies to fit these new norms of slenderness. The obvious interpretation of this development in fashion is that 'the corset which debilitated and inhibited active movement was in effect a physical manifestation of women's forced submission and dependence upon the male' (Davies, 1982, p. 616). This was, however, a very paradoxical submission. It can be argued (Kunzle, 1982) that women also adopted corsets because they were a symbolic statement that they were not pregnant and possibly could not be pregnant. The corset is simultaneously an affirmation of female beauty and a denial of female sexuality. While Mel Davies (1982) supports the view that the corset was an instrument of male oppression, he also argues that the corset reduced the incidence of intercourse, limited exposure to conception and affected gestation and parturition. Davies suggests from medical evidence that the corset caused injury to the cervix often making coitus painful. He also argues that pressure on the abdominal viscera upon the uterus in young women interfered with menstrual flows at puberty, resulting occasionally in amenorrhoea. Finally, he claims that the corset caused uterine problems which made miscarriages and foetal damage more common among women who were corseted. In short, the corset reduced the fertility of middle-class women by comparison with working-class women who were less constrained by corsets. Davies is thus primarily concerned with the demographic implications of the corset, but it is the contradictory symbolic significance of the corset which is most relevant for an analysis of the government of the body. The corset made women conform to a male, middle-class norm of feminine beauty, but it paradoxically reduced and restricted their sexuality by making them less available for coitus. Middle-class men found an outlet for desire among working-class prostitutes. They thus conformed to two norms while being encased within their corsets: (1) the corset offered respectability and beauty; (2) the corset denied desire. Given the contradictory symbolism of the corset, its relationship to twentieth-century anorexia becomes fairly obvious.

The nineteenth-century corset and the twentieth-century fad for slimness through regular dieting and exercise ensure that women conform to certain norms of beauty which are assumed to be attractive to men. In this sense, they illustrate the submissive nature of women in a society organized around patriarchal values and institutions. Apparently women willingly accept these standards through choice because they apparently accept by socialization the notion that slimness is valuable and respectable. The slim body is woman's way into a man's arms, heart and hearth. The corset at least was a necessary condition for success in the marriage market:

> Loving parents now believed that their daughters' chances on the marriage market would be seriously impaired unless they had the correct rigidly upright posture, emaciated bodies, pallid complexion and languid airs, and were prepared to faint at the slightest provocation. The importance attached to these matters was a direct result of the decline of money and the rise of personal choice as the most important factor in the selection of a marriage partner. Girls were now competing with one another in an open market for success in which physical and personal attributes had to a considerable degree taken over the role previously played by the size of the dowry. (L. Stone, 1979, p. 284)

Today's slim woman is less likely to be looking for a marriage partner, much less a permanent one. Slimness is now, under the promotion of the food and drug industry, more geared to the narcissistic ends of personal happiness, social success and social acceptability. The slim body is no longer the product of either an ascetic drive for salvation or of the artificial aid of the corset; it is instead a specific feature of calculating hedonism as the ethic of late capitalism:

> The instrumental strategies which body maintenance demands of the individual resonate with deep-seated features of consumer culture which encourage individuals to negotiate their social relationships and approach their free time activities with a calculating frame of mind. Self preservation depends upon the preservation of the body within a culture in which the body is the passport to all that is good in life. Health, youth, beauty, sex, fitness are the positive attributes which body care can achieve and preserve. (Featherstone, 1982, p. 26)

While slimness may be, for both men and women, the dominant norm of sexual attractiveness, slimness may also be ironically a denial of sexuality or more specifically of procreative functions and fertility. Corsets, jogging and anorexia have one important medical side-effect, namely that they suppress menstruation. We might say that anorexia is over-determined by these contradictory features of femininity. It is an attempt to deny sexuality and thereby to retain a childlike innocence by avoiding menstruation. Anorexics have unhappy and unwanted personal relationships with men, but in becoming slim they adhere to a norm which is assumed to be attractive to men. In the nineteenth century, the corset was the target of feminist reform, which regarded such constraints as against the laws of nature. Fashion was symptomatic of women's social and physical confinement (Leach, 1981). Dieting and exercising in the twentieth century are similarly associated both with a return to a more 'natural' life-style and with the social liberation of women. The right of women to jog in the streets without interference from men is a political right, of their freedom to operate within the public domain. At the same time, jogging and slimming, on the one hand, reduce medical costs and therefore can be regarded as a rationalization of the body in the interests of the state; on the other hand, jogging and slimming increase the sexual attractiveness of women in the interests of consumer culture. Jogging conforms to certain economic and cultural requirements of capitalist society, and it is also associated with a sense of personal freedom on the part of women. Jogging and dieting thus illustrate two themes within Foucault's treatment of knowledge and power. First, at a subjective level they express an enhancement of personal liberty for women. Wearing corsets and jogging are not readily combined, however they are part of a general medicalization of society whereby surveillance and discipline are now self-imposed by the individual. Secondly, it represents a sexualization of society by which we are forced to be sexually acceptable in order to be socially acceptable. However, by becoming desirable we also suppress desire. Dieting was the principal means by which the mediaeval monastic orders controlled the passions in the interests of spirituality. The consumer regime of the modern period simultaneously stimulates and suppresses desire in the interests of increased consumption; the asceticism of diet is harnessed to the hedonism of consumption. The essential cultural contradiction of late capitalism lies here between the asceticism of production (the work ethic) and the hedonism of circulation (the ethic of personal private consumption).

## Women's Complaints

In her study of English madness, Skultans argues that beliefs about the weakness of women and 'about feminine nature are relatively unchanging. For this reason it is not easy to relate beliefs about feminine nature to social and historical changes' (Skultans, 1979, p. 77). In this perspective, it is possible to interpret hysteria, melancholy, menopausal depression and menstrual tension as medical conditions which have 'real' symptoms, but at the same time are ideological constructions which signify the social, rather than the biological, vulnerability of women. Historically, hysteria and melancholy were not simply conditions of women, but specifically of middle-class women. They occurred according to medical opinion because the unmarried wealthy woman was unoccupied and hence prone to nervous disorders which had their physiological origin within the unoccupied womb. The virgin middle-class woman was thus both socially and physiologically 'lazy'. The remedy was marriage and prayer (Turner, 1981).

I have suggested a relationship between the aetiology and symptomatology of anorexia and the effects of tightlacing. Corsets emphasized the vulnerability and weakness of young women. A handbook on child-rearing by William Law in the eighteenth century noted that tightlacing made women 'Poor, pale, sickly, infirm creatures, vapoured through want of spirits' (quoted in L. Stone, 1979, p. 282). Dieting, purging and fasting in the anorexic produces similar sickly symptoms of undernourishment, but these are combined with delusions of fitness, activity and strength. While Skultans may be correct in arguing that it is difficult to relate beliefs about female frailty to social changes, because of their persistent character over many centuries, there are some important changes. Two of these are important for understanding anorexia, namely the commercialization of diet and the arrival of the consumer body. These commercial developments may have the effect of reinforcing the problematic status of women in capitalism. While there are greater opportunities in education and employment for middle-class women in contemporary society than was the case in the 1880s, women are notoriously under-represented in professional occupations. There is thus a contradiction between the achievement orientation within the home and the public restraints on female success outside. My argument is that anorexia belongs to a continuum of body practices which includes dieting, jogging, keep-fit and other forms of secular asceticism. Women's bodies thus become symbolically occupied while remaining economically unoccupied. Given the low fertility of women in the advanced capitalist societies, especially in the middle classes, the imagery of the unoccupied womb and the occupied body is symbolically pertinent to the case of the anorexic woman.

To write in this way is, however, to see the body as a thing, as an object which is the unactive target of social and cultural pressures. The point of Feuerbach's criticism of cognitive rationalism was to assert the indissolubility of the willing, acting, feeling, meaningful, sensuous person. In the case of the anorexic, their sense of the self cannot be detached from their sense of body – they are what they eat and what they do not eat. However, in choosing anorexia they become involved in a paradoxical dialectic which is both social and physiological. Through an act of disobedience, they reproduce the norm of female beauty. Their search for autonomy is fateful, resulting in the dominance of nature over culture. The fact that there are social and collective practices operating on the body of the anorexic should not detract from this political feature of anorexia as a domestic rebellion. Indeed, the general message of this perspective is that 'Addiction, obesity, starvation [anorexia nervosa] are political problems, not psychiatric: each condenses and expresses a contest between the individual and some other person or persons in his environment over the control of the individual's body' (Szasz, 1974, p. 93). To approach anorexia as a political phenomenon points to the intimate and necessary connections between the private and the public domain. In the original theory of patriarchy put forward by Sir Robert Filmer in the sixteenth century, the authority of the monarch was modelled on the authority of husbands over their families and on the theological interpretation of Adam's authority over Eve. Patriarchy as a social system requires this interpretation of private and public authority. Behind this theory of legitimate authority there lies an even more basic argument which is made evident in Richard Sennett's recent study of authority (1980). Power presupposes a dichotomy between reason and desire which corresponds to the public and the private domain. The authority of men over women has been traditionally legitimized as the authority of reason over desire. Anorexia like other 'women's complaints' is part of a symbolic struggle against forms of authority and an attempt to resolve the contradictions of the female self, fractured by the dichotomies of reason and desire, public and private, body and self.

## Calculating Hedonism

While anorexia is often described by psychiatrists as a medical puzzle, it is perhaps less of a puzzle when set within the framework of a late capitalist culture in which narcissism and consumerism are regarded as dominating features (Jacoby, 1980). The post-war increase in real wages, technical improvements in production, improvements in distribution with the development of the department store and mass advertising through television created a vast mass consumer market for personal commodities. These changes were associated with a new personality type which sociologists have referred to as 'the performing self'. The new personality requires validation from audiences through successful performances of the self. The new self is a visible self and the body, suitably decorated and presented, came to symbolize overtly the status of the personal self. Identity became embodied in external performances. Obesity was the new stigma, suggesting sloth, lack of control and hence poor performance (Featherstone, 1982). The characteristics of the narcissistic personality, which sociologists see as especially prevalent amongst the professional middle classes, are as follows: self-love and an inability to form deep, emotional relationships with others; the quest for praise and validation from others; reluctance to involve in relations that are demanding, especially with children or the aged; a horror of ageing and physical deterioration. These are summarized by Lasch (1980, pp. 85–6):

> Chronically bored, restlessly in search of instantaneous intimacy – of emotional titillation without involvement and dependence – the narcissist is promiscuous and often pansexual as well. . . . The bad images he has internalised also make him chronically uneasy about his health, and hypochondria in turn gives him a special affinity for therapy and for therapeutic groups and movement.

To the extent that modern culture can be described as narcissistic in encouraging pseudo-liberation through consumption, therapy groups, the health cult and the norm of happiness, anorexic self-obsession with appearance may be simply an extreme version of modern narcissism. Anorexia is thus a neurotic version of a widespread 'mode of living' which is centred on jogging, keep-fit, healthy diets, weight-watching and calculating hedonism. Looking good and feeling fine are part of the new hedonism which dominates advertisements. While at work something of the traditional Protestant Ethic survives, in the private sphere there is a modern ethic of calculating hedonism. At the same time, the private sphere is increasing an arena of the disciplined body, of which jogging is a primary practice. Like anorexia, jogging is associated with obsessions about food, weight loss and personal control. Like anorexia too, jogging often results in amenorrhoea in women.

The narcissistic culture of modern capitalism is often seen to be evidence of the decline of patriarchal structures in the home and the work-place. Feminism, the decline of the economic centrality of the home as a production unit, the democratization of life-styles and the employment of women are said to have weakened the traditional combination of male authority, gerontocracy, patriarchy and religion. Women under narcissism enjoy a pseudo-liberation from the family, only to be subordinated by the new culture of consumerism. In the case of anorexia, it is easy to argue that one aspect of its aetiology is a male conception of the beautiful woman as thin and athletic. However, anorexia often appears more as a rebellion against parental authority, especially the dominance asserted by the mother. While the particular family structure of the anorexic household can be said to be itself a product of patriarchal relations, these are highly mediated by the particular pattern of socialization characteristic of anorexic aetiology.

# 9
# Disease and Disorder

One central issue in sociology is the idea that human beings are simultaneously part of nature and part of culture. Culture shapes and mediates nature, since what appears as 'natural' in one society is not so in another. It is 'natural' for respectable Japanese to spit in public, but not to blow their noses. Alternatively, we can argue that nature constitutes a limit on human agency, since, as part of a natural environment, we are subject to growth and decay. Reproduction is a requirement for human societies if they are to survive more than one generation. This limiting boundary is of course both uncertain and flexible, because the limits on human 'natural' capacity constantly change. Modern athletes set standards which were assumed to be impossible in previous sporting epochs. More interestingly, genetic engineering is reshaping and redefining what we take 'life' to be.

Most human societies have historically defined this boundary between the human and the inhuman in terms of rituals. For the sake of argument, these rituals may be classified as rituals of inclusion and rituals of exclusion (Hepworth and Turner, 1982). In traditional societies, the fact of birth is not an immediate guarantee of social membership; one has to be transferred from nature to culture by rituals of social inclusion. These typically include religious rituals of initiation: baptism, circumcision and scarification (Brain, 1979). These rituals of inclusion involve cultural work upon the body and their effect is to transform the natural body into a social entity with rights and status. This transformation is brought about by washing, burning and cutting; these transformative interventions are also associated with naming, since having a name is an institutional mark of social membership, but not necessarily of personhood. Being born is not an ultimate guarantee of cultural membership of society, since infanticide was widely practised either implicitly or explicitly in most traditional subsistence societies. Slavery was another alternative and feminist theory claims that women never fully made it across the great divide that separates nature, monsters and unreason from the reality of culture and morality. There are other possibilities. To have a personal name is a good indication that one is a member of a human society. Although chairs and tables are cultural artefacts which embody human labour as commodities, they do not have personal names. Domestic dogs and cats, however, have bodies which have been transformed by human labour (breeding and training) and they are commodities which can be exchanged; they also have personal names and we ascribe character to them. Dogs can be 'neurotic' and 'badly behaved'.

Are domesticated dogs part of nature or culture? Indeed, are they persons? To be born and to be embodied do not in themselves guarantee social membership. The transfer of bodies out of culture back to nature is equally ritualized by exclusionary practices. The dead are buried, cremated or embalmed; their persons are deconstructed by rituals which indicate that they are now to some extent once more 'natural'. Of course, in some cultures a person never dies and may have a capacity for reconstruction at the Day of Judgement. Some persons such as dead saints may continue to have major social roles to play centuries after their physical departure, while most societies have ghosts of one sort or another (Finucane, 1982). It is often argued that advances in modern medicine have made the division between life and death problematic, because the technical definition of death has changed with advances in medical technology (Veatch, 1976). The problem, however, is not simply technical, since there is an essential difference between medical death and 'social death'. Dying is a social process, involving changes in behaviour and a process of assessment which do not necessarily correspond to the physical process of body-death (Sudnow, 1967). Death, like birth, has to be socially organized and, in the modern hospital, is an outcome of team activities. These questions concerning birth, dying, personhood and social membership are indications of a generic issue which hinges on the relationship between nature and culture. The argument of this chapter is that the concept of 'disease' is the most sensitive indicator of the problematic quality of the nature/culture division and that an exploration of the nature of disease provides the best route into the question which lies behind this book as a whole: what is the body?

**Disease versus Illness**

In the philosophy of medicine, there has been considerable debate over the relationship between 'disease' and 'illness' (Agich, 1983). Part of my argument is that the uncertain relationship between these two concepts is a function of the paradoxical and contradictory relationship between nature and culture; the connection between the two is also related to the institutionalization of power and knowledge. We can express the problem initially by examining an expression in everyday language. Although we would have no difficulty with the sentence 'This apple is diseased', we would find it an odd use of the language for somebody to say 'This apple is ill' or 'This apple is sick'. Similarly, we might state that 'His lung is diseased', but rarely 'This person is diseased'. To have a 'diseased mind' is an expression used of some person whose behaviour falls completely outside the pale of normal human activity; such a person is a 'monster'. These everyday examples point to a position in medical philosophy which says that illness is an evaluative concept which is entirely social and practical; disease by contrast is a neutral term referring to a disturbance in an organism or, more technically, to some atypical functional deficiency. This position which sharply distinguishes between illness and disease is best represented by Christopher Boorse (1975). Boorse contrasts 'normativism' and 'functionalism'. Thus, strong normativism argues that all judgements in the medical sciences are evaluative and lack any real descriptive content; weak normativism suggests that health judgements are a mixture of evaluative and descriptive statements. By contrast, functionalism, which Boorse supports, asserts that health and disease may be defined descriptively without reference to values in terms of natural functions which are present within the members of a species. In part, Boorse's argument rests on a distinction between theoretical and practical problems, that is we should not confuse the practical problem of a doctor treating the illness of a patient in a clinic with the theoretical problem of a pathologist analysing a disease in a laboratory context (Feinstein, 1967; Margolis, 1976). On the basis of these arguments, Boorse (1976) came eventually to identify three separate types of 'unhealth'. These are disease, which refers to 'some deviation from a biological norm' (Boorse, 1976 p. 82) illness which is a personal experience of unhealth, and sickness, which is a social role expressing the public dimension of unhealth as in the concept of 'the sick role' (Parsons, 1975).

Normativism – the position that medical judgements are simply evaluative statements – has been dominant in anti-psychiatry and, before turning to a critique of Boorse's defence of functionalism, it is important to consider the problem of 'mental health'. It would, of course, be perfectly possible for a Boorsean functionalist to accept that concepts of mental dysfunction are wholly evaluative while rejecting normativism when applied to diseases of the body. One illustration of normativism as a critique of the concept of 'mental illness' would be labelling theory. Originally used in the sociology of deviance (Gibbs and Erickson, 1975), labelling theory has been extended to explain mental illness as stigmatized behaviour (Scheff, 1974). Psychiatric labels provide an official stamp on behaviour which is regarded as socially unacceptable in the wider society and the effect of these official labels is social exclusion. It is well known, however, that labelling theory cannot genuinely explain the causes of primary deviance; it merely gives a description of labels, stigma and secondary deviance (Gove, 1975). Furthermore, psychiatric labels tend to be used as a last resort when other explanations of deviant behaviour have been exhausted (Morgan, 1975). Although labelling theory has clear weaknesses, it has served a useful function as a critique of the medical model. In particular, labelling theory draws attention to the very different consequences for individual behaviour of different social labels: to call someone 'deviant' has very different consequences from calling them 'sick'. Hence the conversion of deviant categories becomes of special interest historically. Nineteenth-century inebriety was converted into the twentieth-century disease of alcoholism, while homosexuality as a sin was transformed via a disease category to simply a personal preference. Paedophilia may well become a candidate for conversion (B. Taylor, 1981). In psychiatry, diagnostic labels and therapeutic regimes appear to be culturally relativistic and historically variable (Armstrong, 1983). It has proved difficult to locate mental illness within an organic category, since psychological disturbance appears to be a product of life stress and professional categorization (Inglis, 1981). The solution to the problem is to argue (Morgan, 1975) that while diseases can be defined by neutral biological criteria, illness is essentially social since it refers to undesirable deviation from accepted social norms of health and appropriate behaviour. Diseases belong inside nature; illnesses, inside culture. Human beings, because they are ambiguously located in both nature and culture, are subject to both diseases and illnesses. The implication of this position is that we sometimes misdescribe an illness as a disease and the solution is simply to get our categories correct.

We might agree that illness is essentially a cultural phenomenon, but can we agree that disease is simply a fact of nature and not itself subject to cultural processes? The difficulty is that 'disease' is as much contested as 'illness'. The concepts of 'illness' 'disease' and 'health' inevitably involve some judgement which ultimately rests on a criterion of statistical frequency or an ideal state. The 'average individual' does not exist and biological functions can be realized by very different means (Vacha, 1978). Disease is not a fact, but a relationship and the relationship is the product of classificatory processes: 'a disease pattern is a class, or niche in a framework. This framework is a means of approaching or organizing crude experience, that is, for dealing with every-day events in the most satisfactory way' (King, 1954, p. 201). The discovery of a new disease is not, according to this view, epistemologically equivalent to discovering a new butterfly; a new disease is the product of a shift in explanatory frameworks or the identification of a new niche. These changes in framework are linked to changes in institutionalized medicine and to the nature of medical power. The growth of the clinic, for example, meant that

> [t]he whole relationship of signifier to signified, at every level of medical experience, is redistributed: between the symptoms that signify and the disease that is signified, between the description and what is described, between the event and what it prognosticates, between the lesion and the pain that it indicates. (Foucault, 1973, p. xix)

Disease is thus a system of signs which can be read and translated in a variety of ways.

The merit of Foucault's approach to medicine is that it recognizes that changes in the form of knowledge (of disease) are related to forms of power; the weakness of the philosophy of medicine is that it too frequently and too glibly separates the question 'What is disease?' from the question 'What is the function of medical knowledge in the context of medical professionalization?' The language of disease involves judgement as to what is desirable and undesirable, and the medical profession has in modern society enormous institutional purchase on what is to count as the good life. By relegating disease to nature, Boorse denies the impact of cultural values in medicine at all levels. However,

> The conceptual purity which is gained for theoretical health and disease on this account is purchased at the exorbitant price of excusing medicine from the concrete social-cultural world. But it is in this world that disease language functions. (Agich, 1983, p. 38)

The conceptual purity is, in any case, ruled out on the grounds that the natural and cultural realms are interwoven and interlocked. The reality which human beings inhabit is socially constructed and that reality includes biology, which, although a limiting horizon, is still culturally constituted and socially transformed (Berger and Luckmann, 1967). To argue that disease is constituted by classification is thus to raise the question of the ontological status of the body itself.

One implication of Foucault's approach to the relationship between the order of things and the order of words is that the body is itself a cultural object which is the product of classification. This is the conclusion drawn by Armstrong in *The Political Anatomy of the Body*:

> The reality of the body is only established by the observing eye that reads it. The atlas enables the anatomy student . . . to see certain things and ignore others. In effect what the student sees is not the atlas as a representation of the body but the body as a representation of the atlas. (Armstrong, 1983, p. 2)

To approach disease sociologically, we have to combine the notions that (1) disease is a language, (2) the body is a representation and (3) medicine is a political practice. Disease is a social phenomenon, although there may be highly variable individual manifestations of it. All of these problems are, in my view, related to Durkheim's account of collective and individual representations, and we can approach the problem of disease and disorder by, at least initially, considering some contrasts between societies based on mechanical solidarity (where the notion of sacredness is paramount) and those based on organic solidarity (where the dominance of a *conscience collective* is brought into question). The concept of disease as separate from sin and deviance presupposes massive changes in the structure of human societies, namely that they are differentiated. Durkheim's account of the social division of labour is highly pertinent to the conceptualization of disease as a morally neutral entity.

## On the Specialization of Sin, Disease and Deviance

There are two primary strands to Durkheim's sociology and these are the problems of knowledge and order. For Durkheim, the classificatory systems of societies are social facts, that is they have an external and coercive relationship to the individual. Thus, the sacred/profane dichotomy is a classificatory system which compels us to categorize phenomena into a specific division and directs our behaviour in a determinate fashion (Durkheim and Mauss, 1963). It is this sacred language which both interpellates the individual and the society as subjects (Gane, 1983). These classificatory schema have, however, to be related to different forms of moral solidarity within societies and Durkheim proposed a fundamental contrast between societies based on the moral authority of the *conscience collective* and those based on the moral reciprocity of the social division of labour (Durkheim, 1964). In the case of mechanical solidarity, there is no strong sense of individualism and the social division of labour is underdeveloped and minimal. In such a society, social integration is based upon a common culture and the *conscience collective* is embodied in and expressed by common rituals and practices. Society is experienced through common rituals as a sacred entity which has a life and character which stand over the individual; the sacred is experienced and apprehended as massive and extensive. There is also a clear and definite differentiation of sacred and profane realms, which are marked off by common rules, prohibitions and taboos. With the development of the social division of labour, there is a growing sense of the separateness and distinctiveness of the individual from the group as individuals become increasingly specialized and individuated. As a result, the *conscience collective* becomes weaker, attenuated and indeterminate. Religious beliefs and practices become less important in the social integration of societies and the sacred becomes diminished in stature, depth and intensity. Individuals are now bound to society and to each other by reciprocal obligations which emerge out of the economic division of labour.

This transition in social bases of knowledge and order can be illustrated by reference to Durkheim's view of legal change. Law provides a very clear index of social change, since law reproduces the basic forms of social solidarity. The core of legal phenomena is constituted by sanctions and thus the transition from mechanical to organic solidarity can be measured by changes in sanctions. In simple societies, the principal sanctions are repressive; in advanced societies, they are restitutive. Where the *conscience collective* is powerful, infringements against group norms are severely punished by social retribution on the individual offender by violent means. Because the judicial apparatus is minimal, it is society as a whole which exerts repressive retribution on the criminal. The result is that crime reinforces the sense of group solidarity because the crime and the punishment point to and emphasize shared social values. The state is thus the expression of the moral ascendancy of the group over the individual (Hunt, 1978). In societies based on organic solidarity, the law has a very different function; it exists to enforce contracts, to supervise reciprocal relations and to restore imbalances in exchange. Criminal law and repressive sanctions become less central to the integration of society.

As it stands, Durkheim's sociology of law is unsatisfactory and has been critically discussed by both anthropologists and sociologists (Pospisil, 1971; Taylor, Walton and Young, 1973). First, it is not entirely clear that primitive societies have 'law', and secondly, insofar as they do have law, primitive law was far less repressive than Durkheim wanted to argue. The repressive nature of law is more obviously related to the rise of private property and conflicting classes than it is to the dominance of the *conscience collective*. My argument is, however, that much of Durkheim's basic argument can be retained as an ideal typical characterization of social forms and that his thesis can be elaborated and refined to provide an important reflection on the nature of law, religion and medicine. My thesis is that the contemporary division between disease and illness is a feature of the professional and cultural division between law, religion and medicine. In particular, the relationship between social illness and natural disease has to be seen in the context of the secularization of society. In order to present this argument, it is important to consider the differences between three ideal typical societies which, for the sake of convenience, can be called pre-modern, transitional and modern. The nub of the argument is that medicine has replaced religion as the social guardian of morality. This replacement involves a 'medicalization' of the body and society (Zola, 1972).

In pre-modern societies, classificatory distinctions between disease, deviance and sin are either non-existent or underdeveloped. The aetiology of physical disease and social deviance was sought in the moral history and condition of the individual. Health and morality were fundamentally united in practice and in theory. We can thus conceive of these societies as having an outer membrane of protective moral assumptions, a membrane which housed the central sacred core of society. This membrane was periodically attacked by plagues of anti-social facts which had the function of reinforcing the *conscience collective*; these hostile facts were undifferentiated disease–sin–deviance phenomena which constituted a sacrilege. This moral membrane was demarcated by a cluster of rituals of exclusion and inclusion, which maintain the internal purity of the system against external dangers. Punishment of offences against the *conscience collective* was largely repressive and brutal, being exacted on the body of the criminal in a public setting. The scaffold is the symbol of a society grounded in a coherent moral system (Foucault, 1979). Just as there was no clear classificatory discrimination between disease and sin, so there was little professional specialization between lawyers, clergy and doctors. Religious roles embraced a diversity of activities, including welfare, healing and law-finding. Within this type of society, the state did not have extensive social functions and it was not developed as a specialized political instrument. Power was personalized and embodied in the person of the king who had extensive power in theory, but lacked the bureaucratic apparatus to enforce personal authority. In terms of disease categories, epilepsy, venereal disease and leprosy perfectly illustrate the undifferentiated nature of the threats to society, since these conditions were simultaneously religious, moral, medical and legal phenomena.

In a transitional society, there is greater professional specialization both within and between religion, medicine and law. Medical specialists, in particular, attempted to enforce a monopoly over their trade and to exclude unqualified and unlicensed practitioners. In England, an enactment of 1511 attempted to prevent the practice of medicine by persons who were not approved by the appropriate authorities. The difficulties of enforcement are amusingly illustrated by the life of Dr Simon Forman, an Elizabethan doctor–astrologer who had an unlicensed practice among the London elite, much to the chagrin of the Barber–Surgeons and College of Physicians (Rowse, 1974). Although the physicians were organized as a college from 1518, they did not become the Royal College until 1851 and the surgeons did not break away from the United Barber–Surgeons' Company until 1745. Associated with this social division of labour in the 'learned professions' (Carr-Saunders and Wilson, 1933), there was also a mental division of labour in teaching and practice. Sins, crimes and diseases begin to be separated out in the classificatory map of human problems and specialized institutions of control and surveillance – hospitals, asylums and prisons – were developed to deal with specific social problems. In the nineteenth century, hospitals ceased being simply dumping grounds for the poor and were developed as specialized institutions for specific diseases. At the theoretical level, the spiritual and the natural world were increasingly separated under the dominance of scientific naturalism which treated nature as a mechanism, embraced evolutionism and argued that mental phenomena had material causes. The discovery of the tuberculosis bacilli by Pasteur and Koch resulted eventually in the dominance of germ-theory as the principal explanation of disease. Given the existence of microorganisms as the cause of disease, medicine could be presented as a value-free and exact science; the debate about disease could be extracted from theological, social and moral frameworks. With the development of Koch's postulates, every disease had its own specific germ and there was no real need for theological or moral assumptions about diseases in human populations. These changes in knowledge and professional monopolies presuppose changes in the role of the state. A professional control over patients and the ability to exclude competition are only possible with the support of the state and with legal enforcement. Professional specialization and monopolization require the expansion of the state, because practices of social closure which exclude competitors require the coercive support of the state (Parkin, 1979). The professionalization of medicine depends upon coercive acceptance of official definitions of health and disease. These definitions identify the objectivity and reality of disease entities, the facticity of which exists independently of human subjectivity or political judgement (Johnson, 1977).

In modern societies, the public role of religious institutions and professionals is greatly attenuated (Wilson, 1966) and the *conscience collective* becomes effete and indistinct. The moral hinterland of society is worn away by waves of industrialization, urbanization and modernization of consciousness. The remaining moral debris of previous generations is appropriated and dominated by medicine and law. What will count as sin is greatly curtailed in favour of concepts of disease and crime. In turn, what will count as crime or deviance is slowly incorporated by medicine as forms of deviance (alcoholism, malingering, homosexuality and political dissent) are embraced and subsumed under disease. As a result,

> Medicine and the law are the two principal professionalized disciplines of every complex society that have provided an institutionally determinate rule for managing a portion of our prudential interests: the law – in terms of restricting harm or the threat of harm to those interests . . . medicine – in terms of insuring the functional integration of the body (or mind or person), as by care and cure, sufficient for the exercise of our prudential interests. (Margolis, 1976, p. 252)

The state thus regulates the body through the agency of a variety of ideological apparatuses, especially through family law and preventive medicine. The modern medical regimen implies a certain asceticism in morals as the main defence against sexually transmitted diseases, heart disease, stress and cancer. In this sense, religious norms of the good life have been transferred to medicine; the result is that medicine, as an allegedly neutral science of disease, encroaches upon both law and religion, in providing criteria of normality. While the sacral *conscience collective* withers away, medicine provides, as it were, a second-order moral framework – a framework which is, however, masked by the language of disease.

The argument so far has been largely presented in terms of a contrast between three ideal typical characterizations of society, which attempt to provide a sketch of professional specialization and the differentiation of disease, sin and crime. Obviously this typological approach does not attempt to deal in any detail with the complex historical transformation of moral facts from pre-modern to modern society. The sketch simply provides the background to the claim that our conception of disease as a scientific, amoral category is the product of cultural differentiation and professional power. There are without doubt a variety of objections which could be raised against such a background sketch. One objection would be that the relationship between religion and medicine is far more complex than I have suggested. Rather than attempt to defend the argument as a whole, it is interesting to focus briefly on the historical relationship between religion and medicine. The ideal typical outline of pre-modern society suggested that the profane world of disease–sin–crime was undifferentiated, precisely because health and salvation were equated. An objection to such a view would argue that, historically, medicine in the Western tradition has characteristically adhered to a secular conception of disease. In its ancient attempt to establish itself as a profession with a scientific basis, medicine was from Greek times based on natural as opposed to supernatural explanations of health and disease. Thus, Hippocrates rejected supernaturalism in the aetiology of epilepsy in his 'On the sacred disease' (Temkin, 1971). The problem for Greek doctors was to distinguish themselves as clearly as possible from quacks and they did so by claiming a scientific basis to the practice in which illness was seen to have natural causes. Galenic medicine thus came to conceptualize phenomena into three categories (Burns, 1976). The naturals were the functional and structural features of the body, especially the four humours; the non-naturals were basically the environment such as air, food and drink; and the preternaturals were antinature, namely diseases. Health was the proper ordering of the naturals and an appropriate regimen of non-naturals. The role of the doctor was to aid nature in the defeat of preternaturals by manipulating the non-naturals. Supernaturalist causality had a very limited role to play in Greek medical theory and practice. Because the gods had no place in professional Greek medical theory, the implications of Galenism was that the individual had a moral responsibility for the health of his body by following, under the advice of his physician, an appropriate regimen (Temkin, 1973).

This separation of medicine and religion was not invariably categorical and unambiguous, especially in Christian, mediaeval Europe. There was considerable conflict between the religious and the medical view of disease; this conflict had a number of dimensions (Turner, 1981). From the point of view of Christian values, the care and the cure of the sick were acts of charity; the medical fee conflicted with this basic assumption and so professional doctors were seen to be parasitic on the poor and the sick. Furthermore, Christian theodicy could not regard disease as a morally neutral category, since disease was a sign of the health of the soul. This Christian approach to disease faltered on the problems of free will. If God is good, then He cannot be the author of disease, which is written in the moral responsibility of the human being. If God is all powerful, then He must be the author of disease which carries within it a moral lesson. One partial solution to the paradox of medicine and religion was to argue that, although human beings are morally responsible for the diseases which invade them, they are also ultimately responsible to God for the stewardship of their bodies. It was thus in the control of the non-naturals that religious and medical perspectives found an alliance. Stewardship of the body expressed a serious responsibility for the body which the soul inhabited during its earthly existence. The medical regimen and a system of religious stewardship were thus dovetailed as modes of life which promised both longevity and purity. The naturals could be explained scientifically without recourse to supernaturalism or morality, but the realm of the non-naturals was a sphere in which morals, religion and medicine were undifferentiated. The moral regimen and the medical regimen of the body were identical.

## Secularization

This study of the body has presupposed a process of secularization which has transferred the body from an arena of sacred forces to the mundane reality of diet, cosmetics, exercise and preventive medicine. For example, diet was once an aspect of a religious regimen of passions and the aim of asceticism was to liberate the soul from the cloying distractions of desire. In a society where consumption has become a virtue, diet is a method of promoting the capacity for secular enjoyments. Diet was simply one aspect of a more general rationalization of the body; through the application of the natural sciences to the body, the human body became reified and disciplined through systems of gymnastics (Broekhoff, 1972). The division between disease and sin can also be treated as a manifestation of secular intellectualism which extracted the body from its sacral moorings. The body is no longer the focus of a sacred drama involving sacramental ritualism; it has become the object of secular professionalism under the ultimate surveillance of the state. The transition from a sacred canopy to secular surveillance is, however, a great deal more complex than this thesis of rationalization would suggest. Secularization is, in fact, a complex, uneven and contradictory process of cultural change.

In the sociology of religion, there are two distinctive views of secularization which are completely antithetical (Turner, 1983). The first view treats secularization as simply the decline in social significance of organized religion (Wilson, 1982). The church ceases to play a major role in the organization of public life and Christian belief becomes irrelevant in the dominant educational institutions. As a result, religious institutions atrophy and no longer enjoy collective hegemony over the mass of the population. Religion slowly but persistently disappears from the public domain and becomes, insofar as it survives at all, a matter of personal interest. The grip of religion over the body is thus abolished with the spread of secular practices and beliefs. The body is no longer necessarily subject to the transformations of baptismal water, eucharistic feasts or sacramental rites. The body comes now within the gaze of scientific disciplines and institutions which apparently make no assumptions about the supernatural character of implications of our this-worldly materiality.

The alternative argument denies that organized Christianity was a dominant cultural institution in the Middle Ages, because the peasantry fell outside its orthodox mantle. The population was in general indifferent to or ignorant of Christian teaching and practice, adhering to pre-Christian traditions, superstition and pagan practice. The church may have had some control over the urban, literate population, but there was a vast hinterland of opposition to Christian institutions. The popular culture of pre-modern societies was either secular (Burke, 1978) or heretical (Ladurie, 1974). Since there was no 'golden age' of Christianity, there could be no 'decline' of organized, official religion. In this view, the body was not incorporated by Christianity; instead, it was the object of numerous heretical and pagan practices and beliefs. The body was part of the underworld opposition to an urban Christian tradition which attempted to subordinate and to deny the body by orthodox asceticism. However, the sexuality of the ordinary people was largely uncontrolled by and resistant to official attitudes and institutions (Quaife, 1979). The teaching of the church on sexual morality had little effectivity because the cultural apparatus for transmitting and enforcing official norms was relatively weak and underdeveloped (Abercrombie, Hill and Turner, 1980). The human body thus existed within a dense supernatural environment, but this was not essentially Christian. At various points on the way to modern capitalism, the church was able to impose greater control over the population. The Reformation and Counter-Reformation were periods when, through missionary activity, improvements in the quality of the clergy and the growth of literacy, at least some of the pre-Christian pagan culture was swept aside. The Catholic confessional and Protestant conversionism provided the institutional means for a surveillance of everyday sexuality (Hepworth and Turner, 1982). From this perspective, religion does not decline with the growth of capitalism, but rather extends its control over the laity with the growth of systems of mass communication. Secularization is a relatively recent phenomenon and it involves, not so much a steep descent, but an undulating plateau.

As an alternative to both arguments, I want to consider a different view of secularization in which religious functions are simply transferred to secular institutions. Religious beliefs and practices provide the starting point and the model for activities which reappear in secular society under a new garb. Secularization thus involves mutation and reallocation rather than decline and attrition. To some extent, this view of secularization is implicit in Weber's *The Protestant Ethic and the Spirit of Capitalism* (1965). The secular institutions and values of industrial capitalism – calculation, efficiency, hardwork, bureaucratic management, rationalization and vocation – had their origins in the Protestant Reformation. Calvinistic individualism and asceticism were thus transferred through mutation into the secular world of capitalist business. Just as Protestantism transferred the monastery into the family, so secularization pushed religious asceticism onto the factory floor. Protestant rationalism was thus reconstructed and redeployed in capitalist culture as a consequence of a secularization process. A similar argument is implicit in Foucault's commentaries on discipline and sexuality. The discipline of the monastery provided the formative model for the regulation of the body in the school, the army and the factory. Panopticism refined and elaborated the monastic regulations by making them more efficient, effective and rational. Religious disciplines did not decline; they were simply reassembled in a wider and more pervasive social system (Foucault, 1977). Similarly, the Catholic confessional provided a primitive model of surveillance in which the father-confessor intervened in the family between the husband and the wife. The confessional did not deny or silence sexuality, but rendered it expressible and visible. The religious confessional was thus the seed which flowered eventually in a confessing society: 'The confession has spread its effects far and wide. It plays a part in justice, medicine, education, family relationships, and love relations, in the most ordinary affairs of everyday life, and in the most solemn rituals' (Foucault, 1981, p. 59). The assumptions of the transitional confessional – the culture of guilt, the criteria of the true confession, the innocence of talk, and the interior conscience – are through a process of secularization redistributed in a network of modern institutions. The confession did not decline or disappear; it was redeployed in psychoanalysis, police practice, court procedure, modern literature and medicine. In particular, the moral organization of the individual by religious practices is now reallocated to medicine, especially in the intimate relationship between the general practitioner and the family.

## Medical Morality

In eighteenth-century medicine, there was a clear movement at the level of theoretical medicine towards a science of the body through the development of what we might call 'hydraulic mathematics'. The body, for writers like Cheyne, was simply a system of pumps. At the level of medical practice, there was little real change in the medical approach; the equilibrium of the body had to be maintained by appropriate inputs and evacuations. The individual could exercise control over this process through the proper management of the life-process. This management of the individual body had a close relationship to the government of the social body; both required discipline, order and morality. In the final analysis, health depended upon morality, since improper life-styles were the root of personal illness and individual immorality was the product of social disorder. Sickness in the individual was intimately linked with disorder and mismanagement of the social body. Briefly, the wages of sin were disease and death. In this respect, the alliance between doctors and mothers in the control of family morality under the auspices of public health in the nineteenth century represented a continuity with traditional medicine (Donzelot, 1979). Medical prescriptions for health implicitly or explicitly carried with them proscriptions on behaviour which was simultaneously abnormal and immoral. The biologically normal was grounded in notions of socially normal. In the nineteenth century, much of this convergence of medicine and morality was organized around the problem of sexual deviation, especially in women and children. As we have seen, such medical categories as agoraphobia, masturbatory insanity and hysteria were woven into the fabric of Victorian moral notions.

Contemporary medicine is technically sophisticated and claims a scientific basis in the notion that disease is a natural not a cultural phenomenon. Alongside this medical technology, there are also important continuities in the medical evaluation of the contribution of a 'healthy life' to physical well-being. One illustration is the issue of sexually transmitted disease. Gonorrhoea is often described as an 'epidemic' of modern societies, being associated with the liberalization of sexual mores, the increase in homosexuality, the decline in the use of condoms, the increase in asymptomatic carriers, and the resistance of the gonococcal germ to antibiotic treatment. The spread of venereal diseases like gonorrhoea represents an interesting medico-moral problem. In the absence of effective antibiotics gonorrhoea can be controlled by greater medical intervention (such as greater surveillance of carriers and treatment of partners of infected patients), educational programmes recommending more widespread use of male prophylactic devices, or by campaigns against family breakdown, divorce and sexual promiscuity.

The obvious feature of sexually transmitted diseases is that, in most cases, they are associated with illicit or promiscuous sexuality. One implication of medical intervention is thus to make illicit sexuality free from the negative consequences of illness; medical intervention would remove the moral 'lesson' of disease. This paradoxical situation of medicine was evident in the repressive orientation of the National Council for Combating Venereal Disease in the early decades of this century. The Council in Britain warned against the dangers of illicit sexuality, but avoided offering advice on the use of condoms, because this would actually provide an incentive to immoral behaviour (Armstrong, 1983). In more recent times, the paradoxical position of medicine has become evident with respect to herpes, AIDS and cervical cancer. The first two are closely associated with homosexuality, but there is evidence that AIDS can be contracted by a 'naive' subject. Public response to AIDS has been compared to mediaeval attitudes towards leprosy, because AIDS victims have been systematically isolated and excluded by a terrified public. Professional pronouncements have occasionally been equally hysterical; in 1983 the front-page cover of the *Journal of the Australian Medical Association* carried the banner that AIDS was 'the black plague of the eighties'. Although medical research institutes will be funded to discover an antibiotic treatment of modern venereal diseases, one implication of these 'epidemics' is that monogamy, sexual fidelity or celibacy are the primary defences against infection. The same implication is relevant in the case of cervical cancer which is associated with viral infections resulting from promiscuous contacts with males not employing condoms.

Although modern societies possess an extensive body of knowledge relating to the natural causation of disease, it would appear that some diseases are not interpreted in morally neutral terms. Venereal disease is popularly conceptualized as an invasion of the body by alien germs, but the mechanism which, so to speak, opens the sluice-gates permitting nature to invade culture is the deviance of human populations from morality. Since there is concern that the effectivity of antibiotics is in decline, the sluice-gates can be closed by protecting the moral core of society. The sluice-gates are to be controlled by rituals of inclusion and exclusion, because the diseased are not so much 'victims' as 'agents' of a biological disaster.

The illustration of AIDS is, of course, somewhat dramatic and possibly unusual, but it can be taken as indicative of a general alliance between medical practice, morality and sport. Medical standards of appropriate behaviour are now very widespread in modern urban societies in which demographic changes have emphasized the health-issues of geriatric populations. The state has a very clear interest in preventive medicine for promoting health in populations which are rapidly ageing. Personal responsibility for health through exercise, diet and avoidance of drugs, reduces the tax drain of curative medical intervention There is consequently an interesting alliance between the state, the medical profession and the healthy citizen. The monogamous jogger is the responsible citizen, whereas the moral deviant becomes through self-induced illness a burden on the state. The thrust of this argument is that, while we live in a secular society, the traditional *conscience collective* has been to some extent transformed by redistribution within the medical system. These transformations are associated with the decline of organized religion as the principal carrier of the dominant moral ideology, with the professional specialization of law and medicine, and with the role of the state as the guarantor of professional monopolies. The *conscience collective* has diminished and declined, but aspects of its moral content have been transferred into popular concepts of disease as a retribution for an unhealthy life and into the moral world-view of the medical profession.

## Doctors, Women and Sexuality

The argument has been that modern medicine in practice takes a distinctive moral outlook on what is normal behaviour and that such an outlook is not merely accidental but an inevitable feature of medicine. 'Disease' as a category cannot be extricated from a cultural nexus and as a result all judgements about 'normal' functioning carry a moral pay-load. The moral implications of medicine, especially in the area of sexually transmitted diseases, have a more significant impact on women than they do on men. In this respect, the general practitioner has very decisively appropriated the role of the family confessor. Differences in the treatment and attitude of doctors towards men and women are now documented (J.N. Clarke, 1983). In general terms, medicine has been instrumental in the sexist description of women as neurotic and emotional. The myth of female hypochondria supported the idea of women as sick persons and qualified women as patients (Ehrenreich and English, 1978). I take an example from the anonymous and undated *Every Woman's Doctor Book* from the 1930s. The doctor makes a number of paternalistic assumptions about the nature of the normal household and normal sexuality. Sexual intercourse will occur 'once or twice weekly', but not during the woman's period because that would be 'distasteful to people of any delicacy of mind'. The normal marriage will eventually produce children:

> I would advise all young married folk who are actually not in want or ill-health to shoulder the responsibility of a babe early; it may call for some self-denial, a few visits less to places of amusement, a few cigarettes less a week, but the babe will bring such joy and comfort in his tiny hands, that no sacrifice can be weighed against the wonderful gift of his coming.

Interestingly, the first 'babe' is male and the wife is expected to bear the sole burden of his parenting. The conception of children is natural since 'the most pathetic figure in the world is the childless wife'.

It could be argued that this advice belongs to a world which is now moribund and that the medical view of social and individual 'normality' has been radically transformed. There is evidence, however, that doctors continue to regard women in an overtly sexist fashion and this is particularly the case in the area of gynaecology. A general survey of gynaecology texts in the United States published between 1943 and 1973 revealed attitudes of doctors which were consistently condescending and paternalistic (Scully and Bart, 1981). House-bound women are regarded as prone to depressive moods, but these have no real content; women who are liberated from the home through employment or liberated from child-bearing by contraception are regarded as shallow and underdeveloped emotionally:

> The very recent widening of the sphere of feminine activities, with the assumption of the male function of protection and maintenance, has led to a further weakening of the reproductive urge, resulting in the modern 'smart' type – sexless, frigid, self-sufficient. (Scully and Bart, 1981, p. 83)

The assumption behind such an orientation to female patients is that domesticity and mothering are the normal attributes of female biography and that deviations from this normal trajectory result in emotional disturbance – insomnia, depression and migraine (Sacks, 1981). Since the menstrual cycle is often seen as problematic for the woman's emotional stability, the principal solutions are either marriage and pregnancy, or hysterectomy.

Medical advice to women is thus typically based upon taken-for-granted assumptions about normal life-styles for women and these assumptions are ultimately grounded in notions relating to the normality of the nuclear family and the domestic role of women. In this respect, medical morality can be regarded as reactive, regressive and patristic, because there is a radical disjunction between the statistical and prescriptive sense of normality. The position of women has changed in four important respects over the last hundred years and these changes have rendered the medical model of the 'normal' female increasingly obsolete. Women now represent a significant section of the industrial work-force, albeit in the unskilled and casual sector of the market. They have inadequate but important control over reproduction through the availability of contraceptive devices. Women, as a result of legislation relating to marriage, property and divorce, enjoy juridical equality with men in principle. Finally, with the decline of the nuclear family and the growth of single-parent households, women are increasingly likely to assume control of domestic space.

Medical ideology thus assumes a patristic character in making assumptions about women which are sociologically invalid. It is not simply that medical myths qualify and interpellate women as willing patients, but rather that medical assumptions attempt to drive women back into locations which they have to some extent already vacated. The argument is not to be regarded as a conspiracy theory and it is not the case that 'medicalism' is simply forced upon reluctant mothers. A variety of factors reinforce the paternalistic power of doctors over familial relationships. Because social science plays no significant part in the medical curriculum, general practitioners are poorly equipped to understand the social dynamics of illness (O.W. Anderson, 1952; Badgley and Bloom, 1973; Glassner and Freedman, 1979; Pflanz, 1975; R. Strauss, 1957). Women do not occupy commanding positions within the medical profession and a feminist perspective on illness is diluted by professional training: discrimination against women in medical training is well documented and historically well established (Walsh, 1977). The relationship between doctor and patient tends, therefore, to reproduce and reflect the hierarchical relationship between men and women within the household (Gamarnikow, 1978). The general practitioner is thus trained into a conservative and regressive ideology of social relationships, while also inhabiting a social role which has a privileged access to the household and to women in periods of personal crisis. The general practitioner is as a result perfectly equipped to occupy roles of surveillance over the household which husbands can no longer control and monopolize.

## Culture and Disease

Pre-modern societies patrolled their boundaries with dramatic rituals of inclusion and exclusion. As Durkheim observed, the greater the threat from exterior disorders – the crime–sin–disease complex – the greater the sense of group membership and the more prominent the feeling of sacred forces. The threat of extinction from disease and death was vivid and omnipresent, being symbolized in the macabre dance of death and conceptualized in death as the rape of the living. The interior purity of society could be preserved through religious ritualism, especially the sacrament of penance which acted as a religious pump to expurgate sins. Before the rise of an individualistic culture, sins were external and objective, and their origin lay in bodily appetite which undermined reason. The moral apprehension of the individual was implanted in the human species by God and not cultivated by training and moral education (Potts, 1980).

Although I have attempted to argue that where we locate 'disease' is a product of a classificatory scheme which presupposes a nature/culture dichotomy, the concepts of 'culture' and 'civilization' are relatively recent (Elias, 1978). For the *philosophes* of the French Enlightenment, men can be cultivated into civility by appropriate education; society can be transformed via the school. The concepts of *policer* and *civilizer* go together and imply a new social order based on rational ideas. The disorder of society is no longer rooted in the body or objective sin, but in ideas which oppose enlightened cultivation. Desire is to be regulated by training, not ritual and repressive law. People had to be forced to be enlightened, and hence the power of the school to cultivate the child had to have the force of the state to achieve its objectives. The process of compulsory cultivation also required the secularization of society, since it was the church which had traditionally corrupted reason with the mythology of faith. This new power of surveillance over the child required confinement within the protective walls of the family and the school. The disorders of society were now more obviously disorders of the mind rather than the body. As Foucault (1973) has clearly noted, the reforms of the hospital system were also a major legacy of the Enlightenment tradition and the French Revolution. If the citizen had a duty to be rational, he also had an obligation to be well. The sick had a duty to attend the clinic, where knowledge of their disorders could be utilized in the education of the healthy. Medicine was no longer restricted simply to the cure of maladies, but became part of a wider movement in the education of the citizen in the requirements of healthy existence. The political and educational aims of the Enlightenment were to be realized in the medical consultation.

Disease lost its theological aura. It lost its theological significance as a lesson to the sinner and became a natural entity, merely a process in the biological environment, but its moral and social status was somewhat enhanced. Disease can be controlled by social hygiene and by individual education in appropriate life-styles. We can choose to be sick through irrational habits (lack of exercise, abuse, addiction, promiscuity) and these irrational habits are increasingly regarded as deviant. Activities which threaten the health of the individual are also regarded as anti-social and are consequently subject to stigmatization. This stigmatization in everyday relationships has to be seen in a context where the state progressively intervenes in the regulation of behaviour affecting health, for example the control of cigarette advertisements. We can argue, therefore, that, although the theoretical object of medicine is differentiated and secularized, medicine is essentially social medicine, because it is a practice which regulates social activities under the auspices of the state.

It is for this reason that much of the philosophical discussion of the difference between disease and illness is hot-house philosophy; it fails to grasp the historical and social nature of the categories of medicine. 'Disease' has an uncertain status because it lies on the boundaries of 'nature' and 'culture', both of which are social constructs. If 'disease' is an index of the nature/culture relationship, it is also sensitive to gender relations. Diseases are, at least in part, socially distributed along the contours of the social structure; for example, the standardized mortality ratios for diseases of the respiratory system, circulatory system and digestive system are highly correlated with class position. The practice of medicine reflects both the class and gender structure of society. What is 'natural' for women is in everyday life closely monitored by male doctors and this is especially true in issues relating to abortion and pregnancy. For example, the availability of abortion in practice will depend heavily upon a doctor's evaluation of the moral status of the woman (Aitken-Swan, 1977; S. MacIntyre, 1977). It is thus impossible to discuss the nature of 'disease' even in theoretical medicine without locating the concept within a hierarchy of moral evaluations, which in turn have to be understood with reference to power in social groups. 'Disease' is not a unitary concept and not simply a factual statement about natural processes; it is a classification reflecting both material and ideal interests. The importance of such classificatory schemes is that they lead ultimately to questions about the ontological status of the body.

# 10
# Ontology of Difference

To be sure, eating, drinking, and procreation are genuine human functions. In abstraction, however, and separated from the remaining sphere of human activities and turned into final and sole ends, they are animal functions.

K. Marx, *Economic and Philosophical Manuscripts*, 1844

## Introduction

The general argument of this study has been that the social sciences have often neglected the most obvious 'fact' about human beings, namely that they have bodies and they are embodied. When they have taken this factual substratum into account, the results have often been trivial. Sociobiology in particular is a blind alley which suppresses an equally obvious 'fact' about human beings, namely that their biological presence is socially constructed and constituted by communal practices. Although the body has been suppressed as a primary focus in classical social philosophy, there are a number of debates which relate directly to the problem of human embodiment. Two obvious examples would be the mind/body issue in philosophy and the nature/nurture controversy in the sociology of intelligence. More fundamentally, a sociology of the body, by raising the ambiguity of the division between nature and culture, leads to the question: what generically is man? That is, the sociology of the body must ultimately address itself to the nature of social ontology. In sociology, the debate about the nature of being has typically taken a relativistic turn. Since all human attributes appear to be culturally specific, it is difficult to locate any human characteristics which appear to be spatially universal and historically continuous. Sociology appears to bring out the difference between human groups and societies rather than features which unite them. For example, one implication of anthropological research has been that those aspects of sexuality which appear 'natural' are in fact the products of specific cultural arrangements (M. Mead, 1949). If there are no universal features of human nature, then it is difficult to speak of any universal human values. A relativistic view of human attributes is thus conjoined with a relativistic perspective on values. Human beings are just different and, since their 'needs' are relative and socially constructed, we cannot make cross-cultural judgements about whether existing social arrangements can satisfy 'real' needs. Human needs are simply the product of different social arrangements and no judgement can be made as to 'false' or 'real' needs. We cannot postulate that 'needs' exist; we can only observe that human beings talk about having needs. Furthermore, these observations lead to the conclusion that discourses about needs are radically variable, contingent and flexible.

A major alternative to this relativistic view of the body as part of a language by which we organize our experiences socially derives from Marx. It is fairly obvious why Marxism as a radical critique of society should require a stable social ontology, free from so-called bourgeois relativism and subjectivism. Marx wanted to argue that capitalist society was not simply unjust and unequal, but that capitalism had radically disformed and deformed human nature. Furthermore, Marx wanted to adopt a position in which his view of human alienation in capitalist society was not simply a moral opinion or a perspective. The Marxist critique of society had an objective basis, which could be validated by rational enquiry and by an appeal to substantive evidence. Marx's social critique required a universal bed-rock which would show that other interpretations of human needs (such as utilitarianism) were in fact partial, limited and static. Marx wanted, for example, to criticize writers such as J.S. Mill for taking the nature of man in competitive capitalism as a model of the universal species-being. In order to establish this objective critique of capitalism and this rejection of bourgeois models of *homo economicus*, Marx had to depend on some criteria of universal human needs and thus to depend on an articulate social ontology (Soper, 1981). For Marx, nature is never a fixed, external reality, but itself a product of human labour which transforms natural reality into 'humanized nature' and this transformative process also shapes the nature of man as a generic being. This view of the dialectical character of 'man' and 'nature' shaped Marx's whole philosophy and hence there was no radical disjuncture between the early Marx of the *Manuscripts* and the late Marx of *Capital* (Avineri, 1970). It follows that Marx attempted to find a way out of cultural relativism via a theory of philosophical anthropology. Without such a theory of universal human potentiality, Marx could not have bridged the fact/value gap in bourgeois political economy (Lukács, 1971). Social ontology in Marxism is thus not merely a matter of decorative addition, but the essential founding of an objective social critique.

## Marx's Ontology

Despite the centrality of the problem of being to Marx's philosophy of history and society, Marx's account of man-in-nature and nature-in-man has been neglected in Marxist exegesis, especially in the aftermath of Althusser's assault on Marxist humanism (Althusser, 1969). The texts which bear directly on Marx's ontology are somewhat limited in extent (Heller, 1976; Markus, 1978; Mészáros, 1970; Lukács, 1980). This tradition in Marxist exegesis is largely East European, specifically Hungarian, and depends heavily on the influence of Lukács who saw, better than most, the problem of subjectivism in social thought (Lukács, 1974). Various aspects of this Marxist focus on social ontology have found their way into mainstream sociology via the synthetic commentaries of Berger and Luckmann (1967), but the general significance of Marx's ontology for, to give two examples, the sociology of gender and medical sociology, has not been fully appreciated. In Marxism itself, the structuralist interest in the 'scientific' Marx of *Capital* minimized the impact of Marx's ontology which emerged through a debate with Feuerbach and Hegel in the *Economic and Philosophical Manuscripts* of 1844.

Although the implications of Marx's perspective on social ontology are profound, the basic ingredients of his analysis of human nature are relatively direct and coherent. The ingredients of this ontology are incorporated in Marx's account of how social beings transform and appropriate nature through the collective labour process. Man in the generic sense transforms nature which is both the object and condition of his existence through labour which is a conscious, practical social activity, and through this transformation of nature man also continuously and consciously transforms himself. It is important to emphasize two essential features from this preliminary statement. First, human universal characteristics are not fixed and static; what human beings share is not a fixed datum of biology but a universal capacity for transformative labour. Secondly, human capacity to overcome and surmount 'nature' presupposes social relations which enhance human agency and consciousness. These relations are historically variable, depending on the mode of production which is dominant in any given society. It follows that the man–society–nature relationship is essentially and critically historical. One consequence of this transformative capacity is that what Marx called the natural boundary is pushed backwards by the social advance of human potentialities. The limits of nature are reduced by the very advance of social capacities as nature becomes the 'inorganic body' of man (Marx, 1967, p. 293). These early reflections on man's being-in-society have to solve fairly substantial issues in classical social theory: the relationship between man and animals, man and nature, and the ontological leap between nature and society. Marx's answer to these issues is worked around a dialectical analysis of the relationship between the 'naturalization of man' and the 'humanization of nature'. For Marx, the fluidity of the nature/culture dichotomy is carved out by the historical and social character of labour.

Labour in the broad sense of conscious, collective, practical action is the centrepiece of Marx's ontology in that man is a being who has to transform his natural environment through hunting, cultivating crops, domesticating animals and producing the means of production in order to survive. Marx thus describes man as a universal natural being in that man has the capacity and potential to transform any object of the natural world into the subject of his needs. It is in this sense that Marx refers to the naturalization of man in terms of the extension of his 'inorganic body'. At the same time, man imposes a cultural stamp on nature and drives back the natural frontier of external conditions, becoming progressively less dependent on this natural environment. It is this expanding control over nature which both extends human universality and entails the humanization of nature. Productive, intentional activity is the distinctive feature of human intercourse with nature by which man appropriates his environment and at the same time historically forms himself as an active, subjective agent. We cannot, of course, fully understand Marx's ontology without understanding his political economy. In a society based upon the private ownership of the means of production, man no longer labours collectively to satisfy wants through the creation of use-values. In a class society, the means of production are alienated from the working population which is thus forced to sell its labour-power in order to survive. Under these social conditions, the worker is alienated from the product of his labour, alienated from his work, alienated from his species life and fellow men, and finally from nature itself, that is from his inorganic body. Society and nature appear now as external, reified and alien forces controlling the life of the individual rather than as products of human labour itself. Human labour now assumes a negative destructive quality: 'External labour, labour in which man is externalised, is labour of self-sacrifice, of penance. Finally, the external nature of work for the worker appears in the fact that it is not his own but another person's, that in work he does not belong to himself but to someone else' (Marx, 1967, p. 292).

Nature as the site in which human potentiality is realized becomes an alien force as a result of social change under conditions of private ownership and appropriation. Marx's political programme is thus also closely linked with his ontology, since it is only as a result of revolutionary change that man, as a universal being, can repossess his inorganic body. The conceptual proximity of Marx's ontology and political action is clearly borne out in *The German Ideology* where Marx observes that without a revolutionary overthrow of capitalism by communism this universal potentiality of human labour cannot be fully realized and secured (Marx and Engels, 1974, p. 56).

To summarize Marx's argument, the nature of human beings and their being with respect to nature are shaped and constituted by the essential character of men as practical agents who labour collectively to transform their conditions and their own character. The transformation of nature and the development of the human species occur under definite social and historical circumstances. The basic features of this position were expressed by Marx with his usual directness and brevity:

> Men can be distinguished from animals by consciousness, by religion or anything else you like. They themselves begin to distinguish themselves from animals as soon as they begin to *produce* their means of subsistence, a step which is conditioned by their physical organisation. By producing their means of subsistence men are indirectly producing their actual material life. (Marx and Engels, 1974, p. 42)

Nature is not a thing-in-itself, but an extension of man – the inorganic body of human agents – and nature becomes a thing-for-man. Nature exists as an external, objective reality, but it is also transformed by labour and socially appropriated, becoming an internal reality of human development.

## The Body in Nature and Nature in the Body

The fact of human embodiment was a crucial feature of Marx's view of the essence of human nature and the character of human labour. The existence of man was, for Marx, inescapably sensuous. Much of Marx's criticisms of Hegelian idealism was based on the argument that idealism understated or ignored the necessarily sensuous character of human activity. According to the first thesis on Feuerbach, both idealism and materialism had suppressed the conscious, sensuous and active nature of human *praxis*. Idealism grasped the subjective consciousness of human existence but neglected the way in which existence is rooted in sensuous production; materialism grasped man's location in nature, but converted man into a mere machine responding to external pressures (Rotenstreich, 1965). As we have seen, the concept of nature as a world of physical objects independent of man and the concept of man as a thing-like phenomenon (a machine, an hydraulic pump, or as a cog within a clock) both emerged at a specific point in history, namely with the growth of commodity production within a fully monetarized economy (Sohn-Rethel, 1978). Marx, by contrast, regarded both man and nature as the sensuous products of historical and social processes. Although Marx himself, especially in the manuscripts of 1844 and in *The German Ideology*, constantly emphasized this practical, sensuous character of human activity, the fact of human embodiment has not been adequately discussed in recent commentaries on Marx's ontology. Embodiment is a necessary condition of man's sensuous appropriation of nature; embodiment is a precondition for practice. Marxists have not really attempted to conceptualize this rather obvious fact that human sensuous agents require embodiment in order to express their agency. Marxists can, therefore, be criticized alongside sociologists because they 'tend to ignore the body and to "desomatize" social relationships' (Freund, 1982, p. 19). This is an important criticism, but it is possible to develop Marx's ontology in a fruitful and constructive fashion to incorporate the notion of human embodiment. The complexity of the body as both a natural phenomenon and a social product can be exposed by attempting to extend Marx's notion of alienation into a discussion of disease. The problem of disease in the human body in turn brings out the subjective and objective experiences of embodiment.

To repeat a paradox which has formed much of the thematic unity of this study in the sociology of the body, human beings both have and are bodies. Insofar as I have a body, I share a number of characteristics in common with other primates which can be regarded as biological systems and in this sense my body is a natural environment over which I exercise control, but which also exercises restraints over me. Like other phenomena in the environment,

I can touch, feel, smell and see my body. However, I require my body in order to carry out this touching, feeling, smelling and seeing. In exercising control through embodiment, I have immediate and first-order possession over my body in a way which I do not experience with respect to other objects. I possess my body, but there is a sense also in which it possesses me, since the demise of my body is also (at least for all practical purposes) my demise. This embodiment, however, is fundamentally social, since my ontology is necessarily social. Thus, references to the possession of my body do not imply any methodological individualism. To employ a distinction which is now common in the analysis of class positions in Marxism, human beings typically have possession of their bodies, but they do not necessarily have ownership. Although Marx always used the word 'man' in a generic sense, it is also well known that his analysis of the 'human essence' very rarely offered an analysis of the location of women in society and history. One fundamental feature of human society has been that, although women have a phenomenological possession of their bodies, they have rarely exercised full ownership. In this sense, the sexual division of labour has always expressed a fundamental alienation of the body. When Marx defined human ontology in terms of sensuous agency, his definition has to presuppose that human beings enjoy both possession and ownership of their bodies. For Marx, this prerequisite for *praxis* was primarily negated by the emergence of capitalist society in which people are forced to sell their labour given the particular character of the social relations of production. However, the negation of the ownership of our bodies is not specific to capitalism.

Under slavery and patriarchy, ownership of bodies is precluded by the political and legal system of control, so that agents experience their bodies as objects which are ruled externally. There are, of course, various institutional arrangements for the commodification of bodies, and prostitution is notoriously the most ancient of such arrangements. In this respect, prostitution can be regarded as the conversion of a natural asset as a use-value into an exchange-value; under some circumstances, casual homosexuality might also be regarded as a commodification of sexuality, devoid of subjective commitment and affective attachment. This loss of sensuous ownership of the body could, therefore, be regarded as one form of corporeal alienation, since at least one dimension of Marx's use of alienation as a concept involves loss of personal control. In a more interesting fashion, it could also be argued that disease involves a loss of bodily ownership and that disease which entails a loss of self is the most proximate and universal form of human estrangement.

A disease can be regarded as an invasion or at least an unwanted alteration of metabolism which has the consequence of disturbing or curbing my everyday social relations and activity. A disease places a limit or restraint on my creative, sensuous practice. To take one example, gout is a fairly common disease in middle-aged men; it is partly hereditary, but is also associated with poor diet, lack of exercise and alcoholism. Gout is, therefore, probably widespread in the academic community. The immediate cause of gout is an accumulation of uric acid in the blood and the site of the attack is typically the large toe, where the victim feels a sharp and agonizing pain. The agony is usually unexpected and arrives without prior warning. The disease thus has all the features of an uncontrolled invasion of the body as a natural environment. In Greek medicine, gout was called podagra, that is foot-attack, and hence Hippocrates referred to it as the unwalkable disease (Hippocrates, 1886). Podagra is a disease you cannot walk with or upon. Gout is, of course, not simply an invasion of the feet and it is not confined to man, being common in parakeets, turkeys and chickens. Gout thus exists as a metabolic malady in man's domesticated animal environment and also in his internal environment. The tophi which are thus discovered in victims of gout are experienced as an alien intrusion and these tophi, from a phenomenological point of view, indicate the thing-like quality of the body as an environment of the person.

Although disease in this sense is alienation, the important feature of human *praxis* is that even disease can be appropriated and transformed into culture. Gout can also become part of the ensignia and stigmata of personality, since part of the individuality of a person can be known from their gait. Gout in the foot is thus transferred to the personality, which itself becomes gouty, denoting a special type of person. Although gout is clearly very painful as an alien intervention, it also has a certain honorific status as the complaint of the wealthy and the immobile. James Russell Lowell, who suffered much from the malady, referred to gout as a 'handsome complaint' and associated it with persons who enjoyed 'easy circumstances' (Norton, 1894). William Cullen who was Professor of Medicine at Edinburgh in the second half of the eighteenth century advanced the view that gout was an affliction of the intelligent and associated it with an abundance of mental and physical abilities (Donovan, 1975). Cullen's treatment involved diet and an abstemious life-style (Talbott, 1964). Gout like melancholy was a disease of affluence, leisure and urban civilization. The social and metaphorical associations of podagra as a malady of the leisure class are in this respect interesting. Podagra is a malady of immobility; it is both an effect and a cause of stationariness as the disease of 'unwalkability'. Gout thus creates leisureliness (however painfully enforced) and is the badge of leisure. Although it is an alien attack on the extremity of the skeleton, it is also, at least in eighteenth-century culture, appropriated by people as part of their personality and social status. Gout in this respect becomes part of the total self identity and it becomes perfectly meaningful to then refer to 'the gouty individual'. In social terms, diseases are ranked upon a scale of prestige; gout, TB, melancholy and hypertension can be part of the social marks of intelligence, sensitivity and wit.

The point of this argument about gout is that a disease is a cultural paradox. It appears to be, so to speak, in nature but it is also inevitably and deeply social. Gout as a malady brought about by uric acid must make a statement about walking or its absence. Walking is a statement about our social and individual character. The horse and the motor car make walking unnecessary for the leisure class who then take up jogging to avoid the unpleasant side-effects of stationariness. The problem of walking was a starting point for many of Georg Groddeck's illustrations of the psychodynamics of disease and illness in *The Book of the It* (1950); Groddeck was the first follower of Freud to ask systematically 'what is the meaning of illness?' Groddeck denied that health and illness are opposites because both are creations of the organism and they have multiple and contradictory meanings. Groddeck offered the illustration of a person walking from the bathroom, falling and breaking the lower thigh. To walk is to be upright, both physically and socially. Children cannot walk at an early age and are not morally and socially responsible for their actions. To be prostrate is to be helpless, but it is also a cry for help and a confession of the need for help. For Groddeck, there is an intentionality about illness, although it is often hidden from the agent himself. The role of psychoanalysis is in part to provide the interpretation of illness to the victim, so that the sufferer can understand the positive, eufunctional and protective aspect of disease and illness. Illness also expresses human creativity and this grasp of the artistry of illness is perhaps nowhere more beautifully outlined than in Oliver Sacks's study of *Migraine* (1981).

To understand the point of Sacks's commentary on the meaning of migraine, it is useful to adopt Marx's dichotomy of base and superstructure to the world of human illness. We might argue that disease is a malady of the base, that is of the organism which all human beings share by virtue of their location as phenomena in a natural world. Human creativity at an individual level occurs in the superstructure, that is in the social, ideological and moral interpretations they elaborate in response to changes in the organic base. Each disease has an organic grammar, but the speech of the sick patient is highly variable, creative and idiosyncratic. Migraine is something people have, but also something they do. We speak of a 'migraine attack' employing military analogies to suggest an external invasion of the person, but we can also think in terms of migraine behaviour as the activity of a migrainous person. To quote at some length from Sacks,

> If the foundations of migraine are based on universal adaptive reactions, its superstructure may be constructed differently by every patient, in accordance with his needs and symbols.
>
> Thus we can now answer, in principle, the dilemma posed earlier, as to whether migraine is innate or acquired. It is both: in its fixed and generic attributes it is innate, and in its variable and specific attributes it is acquired. . . .
>
> Walking, at its most elementary, is a spinal reflex, but is elaborated at higher and higher levels until, finally, we can recognize a man by the way he walks, by his walk. Migraine, similarly, gathers identity from stage to stage, for it starts as a reflex, but can become a creation. (Sacks, 1981, p. 224)

Walking is a capacity of the biological organism, but it is also a human creation and it can be elaborated to include the 'goose-step', the 'march' and 'about-turn' (Mauss, 1979). Walking is rule-following behaviour, but we can know a particular person by his walk or by the absence of a walk. As Groddeck pointed out, my way of walking may be as much a part of my identity as my mode of speech. Indeed 'the walk' is a system of signs so that the stillness of the migrainous person or the limp of the gouty individual is a communication.

The external disease becomes part of culture and personality through appropriation and interpretation. This Groddeckian perspective may appear peculiar, but it is an important corrective to some of the literature on sickness which fails to grasp the contradictory, dialectical nature of suffering. In medical sociology, the symbolic interactionist perspective involves the application of concepts from deviance theory which treats disease and illness as a uniform negation of the self-concept. In this respect, illness can be seen as a process which increasingly restricts social contacts and undermines the coherence of personal identity. Illness creates a sense of dependency on others and on medical technology. For example, patients who are dependent on kidney dialysis have a constant daily reminder of their dependence on machinery (Strauss and Glaser, 1975). The social isolation brought about by chronic illness leads to experiences of being discredited, rejected and devalued. The chronically sick can no longer exercise conscious agency over their circumstances because they are repeatedly reminded of their dependence and they experience themselves as a burden (Charmaz, 1983). The interactionist argument is that illness is a form of deviance and as such illness is subject to stigmatization which results in a devaluation of the self. The maladies of the body become the stigmatization of the person. Although this perspective clearly illustrates the alienation of the patient from himself and from his social environment, it is important to bear in mind that not all illness is stigmatized; some forms of illness, like some forms of deviance, have a social prestige and in a peculiar way are positively evaluated. Furthermore, negative social labels are not necessarily incorporated by either the sick or the deviant; stigmatization only occurs where isolated individuals actually internalize negative labels. Associations for the blind, diabetics, paraplegics and the like attempt to resist negative labelling by offering a more positive image of the life-style of the sufferer. These comments on the interactionist viewpoint are obviously trivial. The most important issue is the complex and contradictory phenomenological relationship between the individual and their disease.

We express our agency in terms of our interpretation and adoption of disease and illness in the sense that the migraine attack becomes *my* migraine and the gouty leg becomes *my* special mode of walking. We can also exercise agency, however, in becoming ill or diseased in the trivial sense that if I fail to take my regular walk, eat a protein-rich diet and consume rich wines, then I may well become a gouty person. My choice over pipe smoking may also contribute to future illness and so agency operates both at the level of interpretation and in the aetiology of disease. Chronic illness does result in a restriction of social contacts, but there is also evidence that loss of rewarding social relations may be causally connected to the onset of disease. The alleged association between cancer and repressed emotions is thus particularly interesting. The notion that cancerous growths are physical signs of repressed feelings can be traced back to both Wilhelm Reich and Georg Groddeck. Reich sought to explain Freud's cancer of the jaw in terms of Freud's unhappy personal life and his repression of emotion. According to Reich, Freud smoked heavily 'because he wanted to say something which never came over his lips' (Reich, 1975, p. 34). Freud had to bite down his emotions which found their outlet in disease and in this sense Freud 'chose' cancer as an alternative mode of expressivity. For Groddeck, disease is representational, pointing to underlying conflicts and tensions; disease became the symbol of repressed desire (Sontag, 1978). The language we use to describe cancer in terms of unsatisfied and controlled desires emerges out of a consumer culture in which to be complete persons we have to consume, to overspend and to satiate desire. Susan Sontag has suggested, therefore, that modern metaphors of cancer are bits of dangerous ideology, because they hold the patient responsible for disease and thus prevent us from grasping the social aetiology of human misery in capitalist society itself. Her argument is clearly powerful. It would be obscene to argue that workers choose asbestosis as a solution to their repressed emotions; however there is evidence that the onset of cancer is associated with suppressed emotions and that massive stress which is unresolved can act as a trigger for neoplasms (Inglis, 1981).

It is now a commonplace that becoming a patient involves a series of choices from accepting that one is 'ill' to doing something about it (McKinlay, 1973). While the notion of choice is compatible with illness behaviour, the idea that conscious agency might be involved in the causation of disease is far more problematic. The involvement of will in physical disease takes us back to the problem of the relationship between nature and culture; the crucial issue raised by Sontag's discussion of the metaphors of illness is ultimately the relationship between language and reality. Is disease as a classificatory system itself socially constructed by decision-making processes in scientific medicine? Is the body itself merely a social phenomenon?

**Nietzsche versus Marx**

For Marx, nature is an objective reality which forms the environment of human beings and the arena in which they satisfy their needs. However, nature becomes less and less significant for human beings who, through collective and productive labour, push back the boundary of natural restrictions. The relationship between people and nature has thus to be seen as essentially social and historical – the relationship being determined by the mode of production by which values are produced. The existence of nature and the requirement that human needs must be satisfied are fundamental to Marx's ontology and in turn his ontology is basic to his social critique of capitalist society. Without these assumptions, Marx would not be able to draw the distinction between false and real needs. Thus, we have the popular argument in modern Marxism that capitalism ensnares the working class in a web of consumption in which advertising plays upon and creates false needs. Mass consumption is disciplined waste which is ultimately destructive of the individual, social relations and the environment (Baudrillard, 1975). Marx's ontology thus drives us back to the issue of what is universal about 'human nature' and whether 'human nature' is falsified under certain social forms of production. In Marxism, the universal human essence does not reside in biology; what is common to human beings is not certain common biological requirements or physiological characteristics. Human essence is capacity and potentiality; human beings labour upon the natural world to transform it in the satisfaction of their needs and in transforming nature they realize their potentiality as sensuous, practical, creative agents. This 'nature' is, however, subverted by social conditions in which human beings are alienated from themselves, productive conditions and from other human beings. Concepts like 'man' and 'nature' cannot be understood as fixed abstractions:

> Man stands in relation with the objects of the external world as the means to satisfy his needs. But men do not begin by standing 'in this theoretical relation with the objects of the external world'. Like all animals they begin by eating, drinking, etc., i.e. they do not stand in any relation, but are engaged in activity, appropriate certain objects of the external world by means of their actions, and in this way satisfy their needs (i.e. they begin with production . . .) for men who already live in certain social bonds (this assumption follows necessarily from the existence of language), certain external objects serve to satisfy their needs. (Marx, 1956, p. 355, quoted in Schmidt, 1971, pp. 110–11)

Human beings realize their 'human essence' historically and they constantly transform themselves by continuously transforming nature.

Marx thus attempted to avoid relativism by locating his social critique in social ontology. Although the world is always changing as a result of the intervention of humanity in that world and human beings transform themselves as a consequence of their labour on nature, the social ontology of the human species is universalistic. Social structures change constantly, but there is nothing in my being which is essentially different from that of any other. Once Marx's argument is stated in these terms, it is easy to recognize that his ontology could be rendered in entirely different terms with radically different implications. If human beings transform nature through their labour and if human beings are also part of nature, then the ontology of social beings is historically contingent and socially variable. If 'nature' is itself the product of socio-historical practices, then, by labouring on themselves, 'man' is a contingent, socially constructed phenomenon. Marx did not want to deny the existence of nature as independent, objective reality, but he wanted to reject any notion that human beings were determined by objective laws of nature.

The problem of human agency in relation to nature had been addressed by Marx at an early stage of his intellectual development. It played an important part, for example, in his doctoral thesis on 'The difference between the Democritean and Epicurean philosophy of nature' in 1841. Marx praised Epicurus for holding that appearances are real and for asserting the autonomy of human will, but Epicurus's emphasis on human freedom from immutable natural laws also carried the implication that the objectivity of nature was itself dependent on human will. The attraction of Epicureanism for Marx was its atheistic assertion of human freedom from natural laws, that is the subordination of physics to morality. This rejection of necessity and immutability in favour of human agency may have been connected by both Marx and Engels with the doctrines of Heraclitus, namely that everything is flux and that things are always the union of opposites. Although Heraclitus received only passing mention in the works of Marx, (in volume 1 of *Capital*, for example), Heraclitus's view of reality as always becoming was certainly compatible with Marx's general emphasis on the historicity of natural and social phenomena. In *Anti-Dühring*, Engels (1959) also referred to Heraclitus's analysis of the contradictory, changing nature of reality as a 'naive but intrinsically correct conception'. Human agency versus a concept of nature as determining and immutable was thus a permanent and deep concern of both Marx and Engels, and they attempted to solve the issue by regarding nature as determining and determined. Nature was a condition and a requirement of human action in circumstances which are historically changing. If phenomena are always changing historically through human agency, how can things be known? If 'Man' is both the subject of and subject to 'Nature', than 'Man' is constantly subject to and the subject of change. How then could we have knowledge of a being who is always becoming? Marx actually makes the problem more specific by arguing that 'Man' is an abstraction, and that we can only meaningfully talk about real, sensuous men in given empirical circumstances. Since actual men and women are always the products of given social circumstances, and are always changing historically as the result of human intervention, it follows that men and women are always changing, because they constantly change themselves. Since men and women are always becoming, how can we know their nature? Marx is thus faced with a classic dilemma of philosophy which was perfectly expressed by an aphorism of Heraclitus: we can never stand in the same river twice. To express this somewhat differently in modern parlance, the problem for Marx was to grasp sameness in a world of difference.

As we have seen, Marx's social ontology was an attempt to recognize the centrality of human agency and universality within a philosophical system which showed that all concepts are historical and can only be grasped in a specific socio-historical context. Having outlined Marx's position, I want to show some differences between Marx and Nietzsche, but also to note at least some similarities. One point in examining Nietzsche's philosophy is that much recent and influential writing on the body, for example that of Michel Foucault, depends heavily on Nietzsche. I also want to show that, while Nietzsche and Marx shared some common presuppositions, the implications of their views of knowledge are obviously different. One useful starting point is to consider Nietzsche's employment of the Heraclitian paradox of being and becoming, a paradox which is central to Nietzsche's epistemology.

Although Nietzsche's epistemology has not been thoroughly debated, it is of considerable interest, given its proximity to much contemporary structuralist theory. One central theme of this epistemology is the limitations of our knowledge and the ungrounded pretensions of positivism as a system of thought. First, he notes that since the world is flux, it is difficult to accept that we can know 'being' at all. In a world of becoming, we can know that part of being which we have constructed:

> A world in a state of becoming could not in a strict sense be 'comprehended' or 'known'; only in so far as the 'comprehending' and 'knowing' intellect discovers a crude ready-made world put together out of nothing but appearances, but appearances which, to the extent that they are of the kind that have preserved life, have become firm – only to this extent is there anything like 'knowledge', i.e. a measuring of earlier and later errors by one another. (Nietzsche, 1968, sec. 520)

We know that which we have imposed on becoming and that which we have constructed out of the chaotic flux of experiences. It follows secondly that language is a crucial feature of Nietzsche's epistemology of becoming. Since we are forced to think in words, language is the matrix by which we systematize and fix becoming. In section twenty of *Beyond Good and Evil* (1973), Nietzsche noted that it is hardly surprising that there should be strong family resemblances between Indian, Greek and German philosophy. These family resemblances are the product of a language affinity; we are forced to think in similar ways because of an underlying common grammar. Because our grammar involves the subject–predicate relationship as a principal feature, our systematization of reality is inevitably egocentric – I think, I feel, I do. However, these causal notions and egocentric features are products of language (the world of being). The third crucial feature of Nietzsche's epistemology is the emphasis on the usefulness of knowledge in social survival. Language and knowledge grow out of the social need for communication:

> Consciousness does not really belong to man's individual existence but rather to his social or herd nature. . . . We simply lack any organ for knowledge, for 'truth': we 'know' (or believe or imagine) just as much as may be *useful* in the interests of the human herd, the species. (Nietzsche, 1974, sec. 354)

The nature and functions of knowledge are thus located in the evolutionary development and needs of the species, especially in the necessary features of social communication. Language is a requirement of human survival and is rooted ultimately in the physiological basis of human existence. It is relatively easy to grasp how Nietzsche's epistemological theories have provided one aspect of the philosophical background of modern French structuralism.

The immediate background of structuralist thought was provided by the analysis of language by writers like Ferdinand de Saussure, Roman

Jacobson and Louis Hjelmslev (Coward and Ellis, 1977). The elements of this theory of language can be briefly stated. First there is the distinction between speech (*parole*) and language (*langue*) in which the latter is defined as a system of signs whose relationships are determined by linguistic rules. Language is thus a system of signs which has a determinate structure. Secondly, there is the distinction between the acoustic component (the signifier) and the conceptual element (the signified) of the sign. There is no necessary relationship or fit between the signifier 'tree' and the concept of tree. The signified is not a thing but the component of a thing. The relationship between signifiers and signifieds is arbitrary; what makes thought coherent is the system of rules which governs the location and function of individual signs within a system of language. In Foucault's terms, the notion that there is a necessary connection between an 'order of things' and an 'order of words' is a metaphysical assumption of Western epistemology (1970). The knowledge we possess about reality is a knowledge of concepts, since 'reality' is itself a concept which has a meaning as a consequence of its opposition to 'unreality'. Thirdly, structuralism argued that language as a system of signs can be studied as an autonomous, self-generating system which is not dependent on the intentions of its speakers. Structuralism would thus be a synchronic analysis of signs and distinguished from conventional diachronic analyses of Marxism. The history of a system of language was irrelevant for an understanding of the logic of knowledge production. In general, therefore, structuralism was primarily concerned with questions of form rather than substance. The meaning of signifiers – 'cat', 'dog', 'fish' – is determined, not by their relationship to reality, but by a system of formal rules, which is governed by rules of correspondence and difference. 'Man' is defined by difference, primarily by 'woman'.

Although the immediate background to this view of language was provided by the work of Saussure, the structuralist assumptions of writers like Foucault are closely related to Nietzsche's epistemology. In particular, what we know is not reality but simply knowledge and the knowledge we have is the product of language which attempts to impose a system on the flux of becoming. What we know is not things-in-themselves, but signifiers of the world-of-becoming. The knowledge which is generated about the world is always an act of will by which human groups attempt to impose a meaning on the world so that knowledge is always the will to truth.

Although there is much in common, therefore, between Nietzsche and Foucault, there is at least one major gap between their perspectives. Because Nietzsche thought that all human existence involved evaluation of its moral significance, there would always be the possibility of a re-evaluation of values. Nietzsche was not a nihilist; indeed he was a specific critic of life-denying philosophy and an advocate of life-affirming beliefs. Because knowledge is relative and a matter of perspective, there can be no authority imposed from above. Neither God nor the state has a privileged monopoly of knowledge. Although the argument that we can never know being appears to be highly pessimistic, it also has the implication that our knowledge of becoming could in principle be reconstructed. By contrast, Foucault offers no such escape because he rules out the possibility of discovering a transcendental subject to discourse which could give life meaning. His analysis of the psychiatric theory of madness is not designed to offer a better practice; his account of the birth of the clinic does not promise an alternative medicine.

Nietzsche's relativism or perspectivalism brought him to hope for a better moral world, because what we know is useful for our survival. Nietzsche's view of knowledge as practical thus provides a link with Marx who in *The German Ideology* defined language as 'practical consciousness': 'Language, like consciousness, only arises from the need, the necessity of intercourse with other men. . . . Consciousness is, therefore, from the very beginning a social product, and remains so as long as men exist at all' (Marx and Engels, 1974, p. 51). In a similar fashion, Nietzsche saw language as the embodiment of consciousness growing out of the need for social communication in the struggle for the continuity of the human species. It is therefore social being that determines consciousness; in contemporary structuralism, it is discourse which determines being. The subject can only know, feel and experience what is permitted by the rules of discourse. This epistemological argument is thus close to Foucault's position that classification as an expression of will or power determines the possibilities of experience.

**The Body and Difference**

From a structuralist perspective, there is a sense in which the body is socially constructed by discourse and our knowledge of it is only made possible by classificatory procedures – 'In this sense the reality of the body is only established by the observing eye that reads it' (Armstrong, 1983, p. 2). The body is not part of given reality, but an effect of our systemization of becoming. There are two problems with this interpretation of the body. First, it is reductionist in the sense that human conscious action is reduced to the effects of discourse. For example, the sexuality of our bodies is determined by the rules of a discourse which permits individuals to have one sex but not two (Foucault, 1980b). Although the structuralist tradition has been highly critical of reductionism – such as economism – it involves itself in what might be called discursive reductionism. Human agency is either minimized or ignored. For example, structuralism permits no theoretical space for human resistance to discourse, since we are determined by what we are permitted to know. Secondly, despite all the references to pleasure and desire, a structuralist analysis of the body as the effect of discourses ignores the phenomenology of embodiment. The immediacy of personal sensuous experience of embodiment which is involved in the notion of my body receives scant attention. My authority, possession and occupation of a personalized body through sensuous experience are minimized in favour of an emphasis on the regulatory controls which are exercised from outside by experts who are equipped with coercive knowledge of its needs and functions.

An adequate sociology of the body will have to have a number of necessary features. First, it must embrace some notion of agency which works against any simplistic reductionism. Although the body is an environmental limit over which human beings do not have total control, it is also the case that, through embodiment, they exercise some form of corporeal government. They practice in, on and through their bodies. It is partly because of this claim about agency that the conventional distinction between sickness and disease has been rejected. There is an element of agency in the 'selection' of diseases which are methods of coping with dependency and crisis. Diseases are not simply invasions of an environment by an alien entity and there is a sense in which persons appropriate illnesses as part of their individuality. We refer to 'my gout' or 'my migraine' and it would be odd for somebody to talk about 'its gout' and 'its migraine'. This argument is not to deny that morbidity is highly determined by class, ethnicity and gender, but it is to reject an entirely over-socialized conception of embodied persons.

A theory of the body must, secondly, address itself directly to the dichotomy between nature and culture, since the relationship between these is social, historical and contradictory. Bodies are certainly part of nature, but explanation of human behaviour in terms of biology are highly inappropriate. Human practice involves, as Marx noted, the humanization of nature in which nature is appropriated and forced to serve human needs. Our bodies become alienated things only under specific social circumstances. Biology and physiology are themselves classificatory systems which organize and systematize human experience, and they are therefore features of culture not nature. 'Disease' has to be understood as a classification of biological process and, although it would be reasonable for somebody to object that we can die from disease even if it is simply a classification, 'death' is also a variable cultural category being determined by medical decisions-making, medical classification and technological interventions. What will count as a 'living body' or a 'dead body' depends on culture not nature.

Thirdly, an adequate sociology of the body must be social and not individualistic. Such an observation is perhaps too obvious to be worth stating, but much of the phenomenology of embodiment is highly individualistic and fails to recognize fully that personal experience of embodiment is highly mediated by social training, language and social context. My authority over my body may be a necessary feature of human agency, but the nature and extent of that authority depends heavily on my social circumstances – my body may be somebody's legal property.

The sociology of the body thus leads us eventually into the question of social ontology and I have attempted to outline two contrasting positions.

First, there is the tradition of Feuerbach and Marx which approaches the body via a theory of human sensuous practice on nature in which embodiment is social and historical. The realization of human potentiality is only possible under social conditions where human beings are relatively free from external control and alienation. A genuine Marxism is thus not simply about social freedom, but must include some account of the liberation of embodied persons from physical misery. Marx's social ontology leads to a critique of those features of control and constraint over human potentiality which result from social domination. For example, 'women's complaints' are entirely socially produced and it is thus not utopian to consider a society in which they simply would not exist.

The second tradition is associated with the epistemology of Nietzsche and finds its contemporary expression in the work of Foucault. In this structuralist perspective, the body is an effect of complex processes of knowledge which arise from the will to truth. The body arises as the consequence of modern rationalism and it is situated in the context of political struggles which seek to regulate human beings within an administered society. The body is that which is signified by biological, physiological, medical and demographic discourses; it is thus a concept which is the effect of knowledge/power. At one level, such a perspective could have a radical implication, since what has been constructed by knowledge/power could be deconstructed by resistance, but Foucault does not permit any genuine notion of progress. We can illustrate this negative side of Foucault by considering the discussion of sexuality, especially from *Herculine Barbin* (Foucault, 1980b).

Foucault wants to argue correctly that the difference between 'man' and 'woman' is the effect of discourses and is not a natural and obvious fact of biology. The difference is brought about by the impossibility of having two sexes within one body. Modern medical discourse precludes the 'error' of the hermaphrodite by insisting that all bodies are either male or female:

> Biological theories of sexuality, juridical conceptions of the individual, forms of administrative control in modern nations, led little by little to rejecting the idea of a mixture of the two sexes in a single body, and consequently to limiting the free choice of indeterminate individuals. (Foucault, 1980b, p. viii)

However, it could be argued that gay liberation, the feminist movement, the liberalization of laws relating to sex offences and the availability of genetic engineering have blurred the difference between male and female sexuality. There is now a greater awareness that biological sex, sexual personality, sex training, gender identity and sex roles may vary independently. It could also be argued that rigid male/female differences are no longer relevant to late capitalism where technological and social changes have, for example, made 'heavy work' in previously male occupations increasingly insignificant. Citizenship rights in late capitalism are formally universalistic and progressively indifferent to the particularities of biological sex, and insofar as this is the case the ontology of sex difference becomes redundant.

Although Foucault wants to argue that the imposition of difference by discourse represents an extension of power over bodies by an elaboration of knowledge, his pessimism prevents him from recognizing that the possible disappearance of sex difference is an extension of individual rights and potentialities. There is in fact only one line of argument which would be compatible with Foucault's general position, namely that the disappearance of difference represents an expansion of administrative control over persons by repressing their individuality and their sexual uniqueness. Since both 'sameness' and 'difference' are compatible with bureaucratic and localized practices of power over bodies, there is no way out of the iron cage. Although Foucault often appears as a major critic of Cartesian rationalism and although his genealogy appears as a critical weapon, it is difficult to see how there is any real space for agency and resistance. For Foucault, the movement for sexual liberation is based on the false premise that power is always repressive, but he offers no real alternatives since both power and opposition involve discourses. Gender is merely a construction of socio-medical discourses, but so is everything else. Nietzsche's programme for a route out of our dilemma by a principle of the life-affirming instinct is thus largely missing from the work of Foucault. Embodiment is more than conceptual; it is also potentiality and the realization of that potency requires a social critique which recognizes that some societies are more free than others.

**Body Paradoxes**

Sociological theory can be said to be organized around a number of perennial contrasts – agency and structure, individual and society, nature and culture, mind and body. Solutions to these contrasts – voluntarism and determinism – are simultaneously premature and lop-sided, because the contradictions are theoretically creative and productive. We can exercise agency, but we do so in the context of massive structural restraints. We are individuals, but our individuality is socially produced. Human beings as organic systems are part of nature, but their natural environment is also the product of historical practices. 'Nature' is also a product of culture. We are conscious beings, but that consciousness can only be realized through embodiment. The importance of the sociology of the body is that it lies at the axis of these theoretical tensions and it is thus a necessary component of any genuine sociology. The difficulty of providing a coherent account of what we mean by 'the body' is an effect of these theoretic problems.

In order to locate these problems within a theoretically fruitful framework, it has been argued that every mode of production has a mode of desire. Societies exist insofar as they reproduce their means of existence and reproduce their human members. The relations of production in every society presuppose relations of desire; sexual production is never unregulated. The sexual means of reproduction are only biological in the most trivial sense: sexuality is distributed through the society by social relations of possession and ownership, and these relations determine who is to be constituted as appropriately sexed persons and determine what sexual unions are legitimate and desirable. To use the modern jargon, sexuality is inscribed on persons not by the inner discourse of their physiology, but by the exterior discourse of sexual ideology. Every mode of desire thus has its appropriate interpellative ideologies which constitute out of the raw material of human flesh, persons with appropriate sexuality, sexual identity, genders and personalities. These interpellative discourses not only constitute persons with sexualized bodies but insert such agents into the social locations of reproduction (brothels, families, kinship groups and tribes). The mode of production and the mode of desire are focused on the dimensions of individual bodies and populations, and together they generate the four-fold problem of order – reproduction, restraint, representation and regulation – which I have claimed is universal to social formations. Since the control of bodies is essentially the control of female bodies, the sociology of the body has to be simultaneously an analysis of patriarchy and gerontocracy. Discourses of sexual difference, the organization of power through patriarchal relations and the domination of generations by gerontocracy within the household represent essential institutions for the distribution of bodies and the conservation of property. Societies based on private property are necessarily class societies and the precise operation of these forms of control and regulation will vary between classes. Following an earlier argument (Abercrombie, Hill and Turner, 1980), it may well be the case that the organization of such societies depends more on the coherence of the dominant class than the incorporation of subordinate classes. The sexual deviance of the peasantry may have been of concern to the priesthood, but the socio-economic system did not depend on peasant morality. Since different modes of desire correspond to different modes of production, I have argued that some diseases are symbolic of the form of dependency and domination which are constituted by different forms of society. Hysteria, anorexia, onanism and agoraphobia are 'diseases' of sexual dependency, and the language of these complaints is essentially political.

Although every society has a government of the body, there is always resistance and protest. To identify a regimen of the body is not to assume its effectivity and, from the point of view of rulership, it is not surprising that disorders of the social body should be conceptualized in terms of diseases of the individual, especially female, body. Michel Foucault is correct in suggesting that the control of the body is the site of all controls, but the problem with structuralism is that it represents what might be called 'discourse determinism'. By denying subjectivity and embodiment, 'discourse determinism' fails to provide an adequate phenomenology of the body and abandons the idea of the body as sensuous potentiality. I have attempted to illustrate these ideas by emphasizing the importance of human agency and consciousness even in the aetiology of disease. This agency is indicated by the language of ownership by which we indicate our embodiment in the world – my body and not its body, my disease and not its disease, my pain and not its pain. This reproduction of structures by the mode of desire has to operate through my embodiment and the necessity of the reproduction of the system is brought about by the pleasures of copulation. Although various social institutions – the mediaeval church and the communist party – have attempted to separate reproduction and pleasure, the attempt is doomed to failure. This paradoxical fact is a necessary point of resistance.

One peculiar feature of the mode of production in late capitalist society is, however, that it does not require an ascetic mode of desire; indeed, pleasures are produced by the process of commodification and elaborated by the circuit of consumption. The regimen of bodies is no longer based on a principle of ascetic restraint, but on hedonistic calculation and the amplification of desire. Asceticism has been transformed into practices which promote the body in the interests of commercial sensualism. The new anti-Protestant ethic defines premature ageing, obesity and unfitness as sins of the flesh, but this is not an argument against Weber for whom, although capitalism might enhance material wealth and produce pleasure, it could not provide purpose. The problem for capitalism is that the business cycle constantly expands expectations of pleasure which it can never satisfy either continuously or universally. Unlike feudalism, capitalism is not forced to deny satisfactions in the interest of primitive accumulation; it generates permanent expectations of personal gratification which cannot be satisfied within the periodic crises of the circuits of capital.

The pleasures of the body are never wholly incorporated by consumerism; they may become features of individualistic protest and opposition. In late capitalism, there appears to be a radical disjunction between production and desire, which generates widespread dissatisfaction and disillusion. The economic crisis of late capitalism has thus given a new twist to the Malthusian paradox: while human pleasures, being variable, grow by a geometric ratio, the capacity for consumption grows by an arithmetic ratio. The gap between expectation and consumption represents a level of relative deprivation which is the focus of social disequilibrium. It is for this reason that the government of society requires a government of the body. All government involves regulation and regulation is the imposition of uniformities of standard. The bureaucratic regulation of populations takes place, as both Weber and Foucault realized, through the individuation of bodies and in contemporary societies the moral regulation of bodies is brought about under the auspices of health. The problem is that health, like pleasure, is personal and particular, precisely because embodiment is always unique. Bodies may be governed, but embodiment is the phenomenological basis of individuality. The last word, therefore, belongs to Nietzsche:

> For there is no health as such, and all attempts to define a thing that way have been wretched failures. Even the determination of what is healthy for your *body* depends on your goal, your horizon, your energies, your impulses, your errors, and above all on the ideals and phantasms of your soul. Thus there are innumerable

healths of the body; and the more we allow the unique and incomparable to raise its head again, and the more we abjure the dogma of the 'equality of men', the more must the concept of a *normal* health, along with a normal diet, and the normal course of an illness, be abandoned by the medical men. (Nietzsche, 1974, sec. 120)

# References

Abercrombie, N., Hill, S. and Turner, B.S. (1980) *The Dominant Ideology Thesis*, London.

Ackernecht, E.H. (1948) 'Anti-contagionism between 1821 and 1867', *Bulletin of the History of Medicine*, 22, pp. 562–93.

Agich, G.J. (1983) 'Disease and value: a rejection of the value-neutrality thesis', *Theoretical Medicine*, 4, pp. 27–41.

Aitken-Swan, J. (1977) *Fertility Control and the Medical Profession*, London.

Alexander, J. (1982) 'Theoretical logic in sociology', *Positivism, Presuppositions and Current Controversies*, London.

Allegro, J.M. (1964) *The Dead Sea Scrolls: A Reappraisal*, Harmondsworth.

Allegro, J.M. (1968) *Discoveries in the Judean Desert*, Oxford.

Allegro, J.M. (1979) *The Dead Sea Scrolls and the Christian Myth*, Newton Abbot.

Althusser, L. (1969) *For Marx*, London.

Althusser, L. and Balibar, E. (1970) *Reading Capital*, London.

Anderson, O.W. (1952) 'The sociologist and medicine', *Social Forces*, 31, pp. 38–42.

Anderson, P. (1974) *Lineages of the Absolutist State*, London.

Andorka, R. (1978) *Determinants of Fertility in Advanced Societies*, London.

Arendt, H. (1959) *The Human Condition*, New York.

Ariès, P. (1962) *Centuries of Childhood*, London.

Ariès, P. (1974) *Western Attitudes to Death, from the Middle Ages to the Present*, Baltimore and London.

Armstrong, D. (1983) *The Political Anatomy of the Body: Medical Knowledge in Britain in the Twentieth Century*, Cambridge.

Arney, W.R. and Bergen, B.J. (1983) 'The anomaly, the chronic patient and the play of medical power', *Sociology of Health and Illness*, 5, pp. 1–24.

Avineri, S. (1970) *The Social and Political Thought of Karl Marx*, Cambridge.

Badgley, R. and Bloom, S. (1973) 'Behavioral sciences and medical education: the case of sociology', *Social Science and Medicine*, 14, pp. 348–62.

Bakan, D. (1974) 'Paternity in the Judeo-Christian tradition', in A. Eister (ed.), *Changing Perspectives in the Scientific Study of Religion*, New York, pp. 203–16.

Bakhtin, M. (1968) *Rabelais and his World*, Cambridge, Mass.

Banton, M. (1967) *Race Relations*, London and New York.

Barrett, M. and McIntosh, M. (1982) *The Anti-Social Family*, London.

Barthes, R. (1973) *Mythologies*, London.

Barthes, R. (1977) *Sade/Fourier/Loyola*, London.

Barthes, R. (1982) *A Lover's Discourse: Fragments*, New York.

Baudrillard, J. (1975) *The Mirror of Production*, St Louis.

de Beauvoir, S. (1962) *Must We Burn Sade?*, London.

de Beauvoir, S. (1972) *The Second Sex*, Harmondsworth.

Beecher, J. and Bienvenu, R. (1972) *The Utopian Vision of Charles Fourier, Selected Texts on Work, Love and Passionate Attraction*, London.

Bell, D. (1980) *The Winding Passage; Essays and Sociological Journeys 1960–1980*, Cambridge, Mass.

Bennett, T. (1979) *Formalism and Marxism*, London.

Benoist, J.M. (1978) *The Structuralist Revolution*, London.

Berger, P.L. (1974) 'On the obsolescence of the concept of honour', in P.L. Berger, B. Berger and H. Kellner, *The Homeless Mind*, Harmondsworth, pp. 78–89.

Berger, P.L. and Kellner, H. (1965) 'Arnold Gehlen and the theory of institutions', *Social Research*, 32, pp. 110–15.

Berger, P.L. and Luckmann, T. (1963) 'Sociology of religion and sociology of knowledge', *Sociology and Social Research*, 47, pp. 417–27.

Berger, P.L. and Luckmann, T. (1967) *The Social Construction of Reality: Everything that Passes for Knowledge in Society*, London.

Bernardo, A.S. (1975) 'Petrarch's Laura: the convolutions of a humanistic mind', in R.T. Morewedge (ed.), *The Role of Woman in the Middle Ages*, Albany, NY, pp. 65–89.

Black, H. (1902) *Culture and Restraint*, London.

Black, M. (ed.) (1961) *The Social Theories of Talcott Parsons*, Englewood Cliffs, NJ.

Bloor, M. (1976) 'Bishop Berkeley and the adenotonsillectomy enigma: an exploration of variation in the social construction of medical disposals', *Sociology*, 10, pp. 43–61.

Boardman, P. (1978) *The Worlds of Patrick Geddes*, London.

Bonser, W. (1963) *The Medical Background of Anglo-Saxon England*, London.

Boorse, C. (1975) 'On the distinction between disease and illness', *Philosophy and Public Affairs*, 5, pp. 49–68.

Boorse, C. (1976) 'What a theory of mental health should be', *Journal for the Theory of Social Behaviour*, 6, pp. 61–84.

Borges, J.L. (1973) *A Universal History of Infamy*, Harmondsworth.

Brain, R. (1979) *The Decorated Body*, London.

Braudel, F. (1974) *Capitalism and Material Life 1400–1800*, London.

Breger, L. (1981) *Freud's Unfinished Journey*, Boston and Henley.

Brody, S.N. (1974) *The Disease of the Soul*, Ithaca, NY, and London.

Broekhoff, J. (1972) 'Physical education and the reification of the human body', *Gymnasion*, IX, pp. 4–11.

Brown, N.O. (1966) *Love's Body*, New York.

Bruch, H. (1978) *The Golden Cage: The Enigma of Anorexia Nervosa*, Cambridge.

Bullough, V.L. (1966) *The Development of Medicine as a Profession*, New York.

Burckhardt, J. (1960) *The Civilization of the Renaissance in Italy*, London.

Burke, P. (1978) *Popular Culture in Early Modern Europe*, London.

Burns, C.R. (1976) 'The non-naturals: a paradox in the western conception of health', *Journal of Medicine and Philosophy*, 1, pp. 202–11.

Burton, R. (1927) *The Anatomy of Melancholy*, London.

Canning, J.P. (1980) 'The corporation in the political thought of the Italian jurists of the thirteenth and fourteenth centuries', *History of Political Thought*, 1, pp. 9–32.

Carr-Saunders, A.M. and Wilson, P.A. (1933) *The Professions*, London.

Carter, A. (1979) *The Sadeian Woman: An Exercise in Cultural History*, London.

Charcot, J.M. (1889) *Disorders of the Nervous System*, London.

Charmaz, K. (1983) 'Loss of self: a fundamental form of suffering in the chronically ill', *Sociology of Health and Illness*, 5, pp. 168–95.

Chenu, M.D. (1969) *L'Eveil de la conscience dans la civilisation médiévale*, Paris.

Chernin, K. (1981) *The Obsession: Reflections on the Tyranny of Slenderness*, New York.

Cherno, M. (1962–3) 'Feuerbach's "Man is what he eats": a rectification', *Journal for the History of Ideas*, 23–4, pp. 397–406.

Cheyne, G. (1724) *Essay on Health and Long Life*, London.

Cheyne, G. (1733) *The English Malady*, London.

Cheyne, G. (1740) *An Essay on Regimen*, London.

Cheyne, G. (1742) *The Natural Method of Curing the Diseases of the Body*, London.

Chua, B.H. (1981) 'Genealogy as sociology? Michel Foucault', *Catalyst*, no. 14, pp. 1–22.

Clarke, J.N. (1983) 'Sexism, feminism and medicalism: a decade review of literature on gender and illness', *Sociology of Health and Illness*, 5, pp. 62–82.

Clay, R.M. (1909) *The Mediaeval Hospitals of England*, London.

Cockerham, W.C. (1982) *Medical Sociology*, Englewood Cliffs, NJ.

Cooley, C.H. (1964) *Human Nature and the Social Order*, New York.

Cornaro, L. (1776) *Discourses on a Sober and Temperate Life*, London.

Coulter, H.L. (1977) *Divided Legacy: A History of the Schism in Medical Thought*, Washington, 3 vols.

Coward, R. and Ellis, J. (1977) *Language and Materialism: Developments in Semiology and the Theory of the Subject*, London.

Crapanzano, V. (1973) *The Hamadsha: A Study in Moroccan Ethno-psychiatry*, Berkeley, Calif.

Crisp, A.H., Palmer, R.L. and Kalucy, R.S. (1976) 'How common is anorexia nervosa? A prevalence study', *British Journal of Psychiatry*, 128, pp. 549–54.

Crisp, A.H. and Toms, D.A. (1972) 'Primary anorexia or weight phobia in the male: report on 13 cases', *British Medical Journal*, 1, pp. 334–8.

Dahrendorf, R. (1968) *Essays in the Theory of Society*, London.

Danto, A.C. (1975) *Sartre*, London.

Davies, M. (1982) 'Corsets and conception: fashion and demographic trends in the nineteenth century', *Comparative Studies in Society and History*, 24, pp. 611–41.

Davis, N. (1971) 'The reasons of misrule: youth groups in sixteenth-century France', *Past and Present*, 50, pp. 41–75.

Debus, A.G. (ed.) (1972) *Science, Medicine and Society in the Renaissance*, New York.

Deleuze, G. and Guattari, F. (1977) *Anti-Oedipus, Capitalism and Schizophrenia*, New York.

Derrida, J. (1978) *Writing and Difference*, London.

Donovan, A.L. (1975) *Philosophical Chemistry in the Scottish Enlightenment*, Edinburgh.

Donzelot, J. (1979) *The Policing of Families*, New York.

Dossey, L. (1982) *Space, Time and Medicine*, London.

Douglas, M. (1970) *Purity and Danger: An Analysis of Concepts of Pollution and Taboo*, Harmondsworth.

Druss, R. and Silverman, J. (1979) 'The body image and perfectionism of ballerinas', *General Hospital Psychoanalyst*, 1, pp. 115–21.

Duby, G. (1978) *Medieval Marriages: Two Models from Twelfth-century France*, Baltimore and London.

Dupont-Sommer, A. (1961) *The Essene Writings from Qumran*, Oxford.

Durkheim, E. (1961) *The Elementary Forms of the Religious Life*, New York.

Durkheim, E. (1964) *The Division of Labour in Society*, Glencoe, Ill.

Durkheim, E. (1970) *Suicide: A Study in Sociology*, London.

Durkheim, E. and Mauss, M. (1963) *Primitive Classification*, London.

Edelstein, L. (1937) 'Greek medicine in its relation to religion and magic', *Bulletin of the History of Medicine*, 5, pp. 201–46.

Ehrenreich, B. and English, D. (1978) *For Her Own Good: 150 Years of the Expert's Advice to Women*, New York.

Eisenstadt, S.N. (ed.) (1968) *The Protestant Ethic and Modernization*, New York.

Eliade, M. (1958) *Rites and Symbols of Initiation*, New York.

Elias, N. (1978) *The Civilizing Process*, Oxford.

Engelhardt, H.T. (1974) 'The disease of masturbation: values and the concept of disease', *Bulletin of the History of Medicine*, 48, pp. 234–48.

Engels, F. (1934) *Dialectics of Nature*, Moscow.

Engels, F. (1952) *The Condition of the Working Class in England in 1844*, London.

Engels, F. (1959) *Anti-Dühring*, Moscow.

Engels, F. (1976) *Ludwig Feuerbach and the End of Classical Philosophy*, Peking.

Engels, F. (n.d.) *The Origin of the Family, Private Property and the State*, Moscow.

Epstein, I. (1959) *Judaism*, Harmondsworth.

Ewen, S. and Ewen, E. (1982) *Channels of Desire*, New York.

Featherstone, M. (1982) 'The body in consumer culture', *Theory, Culture & Society*, 1, pp. 18–33.

Feighner, J.P. (1972) 'Diagnostic criteria for use in psychiatric research', *Archives of General Psychiatry*, 26, pp. 57–63.

Feinstein, A.R. (1967) *Clinical Judgment*, Baltimore.

Figes, E. (1978) *Patriarchal Attitudes*, London.

Finucane, R.C. (1982) *Appearances of the Dead: A Cultural History of Ghosts*, London.

Fischer, D.H. (1977) *Growing Old in America*, New York.

Flandrin, J.L. (1975) 'Contraception, marriage and sexual relations in the Christian West', in R. Forster and O. Ranum (eds), *Biology of Man in History*, Baltimore and London, pp. 23–47.

Foucault, M. (1967) *Madness and Civilization: A History of Insanity in the Age of Reason*, London.

Foucault, M. (1970) *The Order of Things: An Archaeology of the Human Sciences*, London.

Foucault, M. (1972) *The Archaeology of Knowledge*, London.

Foucault, M. (1973) *The Birth of the Clinic: An Archaeology of Medical Perception*, London.

Foucault, M. (1977) *Language, Counter-Memory, Practice*, Oxford.

Foucault, M. (1979) *Discipline and Punish: The Birth of the Prison*, Harmondsworth.

Foucault, M. (1980a) *Power/Knowledge: Selected Interviews and Other Writings 1972–1977*, Brighton.

Foucault, M. (1980b) *Herculine Barbin: Being the Recently Discovered Memoirs of a Nineteenth-century French Hermaphrodite*, Brighton.

Foucault, M. (1981) *The History of Sexuality*, vol. 1. *An Introduction*, London.

Fox-Davies, A.C. (1909) *A Complete Guide to Heraldry*, London.

Freedman, R. (1975) *The Sociology of Human Fertility*, New York.

Freud, S. (1960) *Totem and Taboo*, London.

Freud, S. (1979) *Civilization and its Discontents*, London.

Freud, S. and Breuer, J. (1974) *Studies in Hysteria*, Harmondsworth.

Freund, P.E.S. (1982) *The Civilized Body: Social Domination, Control and Health*, Philadelphia.

Fromm, E. (1941) *Escape from Freedom*, New York.

Frye, R.M. (1954) 'Swift's Yahoo and the Christian symbols for sin', *Journal of the History of Ideas*, 5, pp. 201–17.

Gamarnikow, E. (1978) 'Sexual division of labour: the case of nursing', in A. Kuhn and A.M. Wolpe (eds), *Feminism and Materialism: Women and Modes of Production*, London, pp. 96–123.

Gane, M. (1983) 'Durkheim: the sacred language', *Economy and Society*, 12, pp. 1–47.

Gardiner, J. (1975) 'Women's domestic labour', *New Left Review*, no. 89, pp. 47–58.

Garfinkel, H. (1956) 'Conditions of successful degradation ceremonies', *American Journal of Sociology*, 61, pp. 420–4.

Geoghegan, V. (1981) *Reason and Eros: The Social Theory of Herbert Marcuse*, London.

Gibbs, J. and Erickson, M. (1975) 'Major developments in the sociological study of deviance', *Annual Review of Sociology*, 1, pp. 21–42.

Glass, D.V. and Eversley, D.E.C. (1965) *Population in History*, Chicago.

Glassner, B. and Freedman, J. (1979) *Clinical Sociology*, New York and London.

Goffman, E. (1968) *Stigma: Notes on the Management of Spoiled Identity*, Harmondsworth.

Goffman, E. (1969) *The Presentation of Self in Everyday Life*, London.

Goffman, E. (1970) *Strategic Interaction*, Oxford.

Goffman, E. (1972) *Interaction Ritual*, London.

Goldmann, L. (1973) *The Philosophy of the Enlightenment*, London.

Gouldner, A.W. (1967) *Enter Plato: Classical Greece and the Origins of Social Theory*, London.

Gouldner, A.W. (1971) *The Coming Crisis of Western Sociology*, London.

Gove, W.R. (ed.) (1975) *The Labelling of Deviance: Evaluating a Perspective*, New York.

Gramsci, A. (1971) *Selections from Prison Notebooks*, London.

Grierson, H.J.C. (1956) *Milton and Wordsworth*, London.

Groddeck, G. (1950) *The Book of the It*, London.

Groddeck, G. (1977) *The Meaning of Illness*, London.

Grylls, D. (1978) *Guardians and Angels: Parents and Children in Nineteenth-century Literature*, London.

Guerra, F. (1969) 'The role of religion in Spanish American medicine', in F.N.L. Poynter (ed.), *Medicine and Culture*, London, pp. 179–88.

Gull, W.W. (1874) 'Anorexia nervosa (apepsia hysterica, anorexia hysteria)', *Transactions of the Clinical Society of London*, 7, p. 22.

Hamilton, P. (1983) *Talcott Parsons*, London.

Hanfi, Z. (1972) *The Fiery Brook: Selected Writings of Ludwig Feuerbach*, Garden City, NY.

Hanna, T. (1970) *Bodies in Revolt: A Primer in Somatic Thinking*, New York.

Harper, P. (1982) *Changing Laws for Changing Families*, Melbourne.

Harré, R. (1964) *Matter and Method*, London.

Heller, A. (1976) *The Theory of Need in Marx*, London.

Heller, A. (1978) *Renaissance Man*, London.

Heller, A. (1979) *A Theory of Feelings*, Assen.

Heller, A. (1982) 'The emotional division of labour between the sexes: perspectives on feminism and socialism', *Thesis Eleven*, no. 5/6, pp. 59–71.

Henderson, L.J. (1935) 'Physician and patient as a social system', *New England Journal of Medicine*, 212, pp. 819–23.

Henderson, L.J. (1936) 'The practice of medicine as applied sociology', *Transactions of the Association of American Physicians*, 51, pp. 8–15.

Hepworth, M. and Turner, B.S. (1982) *Confession: Studies in Deviance and Religion*, London.

Hewson, M.A. (1975) *Giles of Rome and the Medieval Theory of Conception*, London.

Hill, C. (1964) *Society and Puritanism in Pre-Revolutionary England*, London.

Hindess, B. and Hirst, P.Q. (1975) *Pre-Capitalist Modes of Production*, London.

Hippocrates (1886) *The Genuine Works of Hippocrates*, New York, 2 vols.

Horkheimer, M. and Adorno, T. (1973) *Dialectic of Enlightenment*, London.

Howell, M. and Ford, P. (1980) *The True History of the Elephant Man*, Harmondsworth.

Hubert, H. and Mauss, M. (1964) *Sacrifice: Its Nature and Function*, London.

Huby, P. (1969) *Greek Ethics*, London.

Hudson, L. (1982) *Bodies of Knowledge: The Psychological Significance of the Nude in Art*, London.

Hunt, A. (1978) *The Sociological Movement in Law*, London.

Husserl, E. (1978) *The Origin of Geometry: An Introduction*, New York.

Hymer, S. (1972) 'The multinational corporation and the law of uneven development', in J. Bhagwati (ed.), *Economics and World Order from the 1970s to 1990s*, London, pp. 113–140.

Inglis, B. (1981) *The Diseases of Civilization*, London.

Jaeger, W. (1944) *Paideia: The Ideals of Greek Culture*, New York.

Jacoby, R. (1980) 'Narcissism and the crisis of capitalism', *Telos*, 44, pp. 58–65.

James, W. (1929) *The Varieties of Religious Experience*, London.

Janik, A. and Toulmin, S. (1973) *Wittgenstein's Vienna*, London.

Jay, M. ( 1973) *The Dialectical Imaginations: A History of the Frankfurt School and the Institute of Social Research 1923–1950*, London.

Jeffrey, J.B. (1954) *Retail Trading in Britain 1850–1950*, London.

Johnson, T. (1977) 'The professions in the class structure', in R. Scase (ed.), *Industrial Society: Class, Cleavage and Control*, London, pp. 93–110.

Kalucy, R.S., Crisp, A.H. and Harding, B. (1977) 'A study of 56 families with anorexia nervosa', *British Journal of Medical Psychology*, 50, pp. 381–95.

Kamenka, E. (1970) *The Philosophy of Feuerbach*, London.

Kantorowicz, E.H. (1957) *The King's Two Bodies*, Princeton, NJ.

Kavolis, V. (1980) 'Logics of selfhood and modes of order: civilizational structures for individual identities', in R. Robertson and B. Holzner (eds), *Identity and Authority: Exploration in the Theory of Society*, Oxford, pp. 40–60.

Kealey, E.J. (1981) 'Critical theory, commodities and the consumer society', *Theory, Culture & Society*, 1, pp. 66–83.

Kellner, D. (1983) 'Critical theory, commodities and the consumer society', *Theory, Culture & Society*, 1 (3), pp. 66–84.

Kern, S. (1975) *Anatomy and Destiny: A Cultural History of the Human Body*, New York.

King, L.S. (1954) 'What is disease?', *Philosophy of Science*, 21, pp. 193–203.

Kolakowski, L. (1978) *Main Currents of Marxism*, Oxford, 3 vols.

Kolko, G. (1961) 'Max Weber on America: Theory and Evidence', *History and Theory*, 1, pp. 243–60.

Kudlien, F. (1976) 'Medicine as a "Liberal Art" and the question of the physician's income', *Journal of the History of Medicine and Allied Sciences*, 3, pp. 448–59.

Kuhn, A. and Wolpe, A.M. (eds) (1978) *Feminism and Materialism: Women and Modes of Production*, London.

Kunzle, D. (1982) *Fashion and Fetishism*, London.

Kurzweil, E. (1980) *The Age of Structuralism: Lévi-Strauss to Foucault*, New York.

Lacan, J. (1977) *Ecrits: A Selection*, London.

Ladurie, E. Le Roy (1974) *The Peasants of Languedoc*, Urbana, Ill.

Lasch, C. (1979) *The Culture of Narcissism*, New York.

Lasch, C. (1980) 'Recovering reality', *Salmagundi*, no. 42, pp. 44–7.

Laslett, P. (1968) *The World We Have Lost*, London.

Laslett, P. (ed.) (1972) *Household and Family in Past Time*, Cambridge.

Lawrence, M. (1979) 'Anorexia nervosa – the control paradox', *Women's Studies International Quarterly*, 2, pp. 93–101.

Leach, W. (1981) *True Love and Perfect Union: The Feminist Reform of Sex and Society*, London and Henley.

Lefebvre, H. (1971) *Everyday Life in the Modern World*, London.

Leiss, W. (1972) *The Domination of Nature*, New York.

Lemert, G.C. and Gillan, G. (1982) *Michel Foucault: Social Theory as Transgression*, New York.

Lévi-Strauss, C. (1969) *Totemism*, Harmondsworth.

Lévi-Strauss, C. (1970) *The Raw and the Cooked*, London and New York.

Lévi-Strauss, C. (1976) *Tristes Tropiques*, Harmondsworth.

Lewis, C.S. (1936) *The Allegory of Love*, London.

Liu, A. (1979) *Solitaire*, New York.

Lloyd, G.E.R. (1978) *Hippocratic Writings*, Harmondsworth.

Lofland, L.H. (1973) *The World of Strangers: Order and Action in Urban Public Space*, New York.

Lukács, G. (1971) *History and Class Consciousness*, London.

Lukács, G. (1974) *The Destruction of Reason*, London.

Lukács, G. (1980) *The Ontology of Social Being*, London, 3 vols.

McDonough, R. and Harrison, R. (1978) 'Patriarchy and relations of production', in A. Kuhn and A.M. Wolpe (eds), *Feminism and Materialism: Women and Modes of Production*, London, pp. 11–41.

MacIntyre, A. (1970) *Marcuse*, London.

MacIntyre, S. (1977) *Single and Pregnant*, London.

McKinlay, J.B. (1973) 'Social networks, lay consultation and help-seeking behaviour', *Social Forces*, 51, pp. 255–92.

MacKinney, L. (1952) 'Medical ethics and etiquette in the early Middle Ages: the persistence of Hippocratic ideals', *Bulletin of the History of Medicine*, 26, pp. 1–31.

McLachlan, H. and Swales, J.K. (1980) 'Witchcraft and anti-feminism', *Scottish Journal of Sociology*, 4, pp. 141–66.

MacLean, I. (1980) *The Renaissance Notion of Women*, Cambridge.

MacLeod, S. (1981) *The Art of Starvation*, London.

Macpherson, C.B. (1962) *The Political Theory of Possessive Individualism*, Oxford.

Malthus, T.R. (1914) *An Essay on Population*, London, 2 vols.

Mandel, E. (1962) *Marxist Economic Theory*, London.

Mann, M. (1970) 'The social cohesion of liberal democracy', *American Sociological Review*, 35, pp. 423–39.

Mannoni, M. (1973) *The Child, his 'Illness' and the Others*, Harmondsworth.

Marcel, G. (1951) *Le Mystère de l'Etre*, Paris.

Marcuse, H. (1964) *One-Dimensional Man*, London.

Marcuse, H. (1968) *Negations*, London.

Marcuse, H. (1969) *Eros and Civilization*, London.

Margalith, D. (1957) 'The ideal doctor as depicted in ancient Hebrew writings', *Journal of the History of Medicine and Allied Sciences*, 12, pp. 37–41.

Margolis, J. (1976) 'The concept of disease', *Journal of Medicine and Philosophy*, 1, pp. 238–54.

Markus, G. (1978) *Marxism and Anthropology*, Assen.

Marshall, G. (1982) *In Search of the Spirit of Capitalism: An Essay on Max Weber's Protestant Ethic Thesis*, London.

Marx, K. (1926) *The Eighteenth Brumaire of Louis Bonaparte*, London.

Marx, K. (1967) *Writings of the Young Marx on Philosophy and Society*, New York.

Marx, K. (1970) *Economic and Philosophical Manuscripts of 1844*, London.

Marx, K. (1974) *Capital*, London, 3 vols.

Marx, K. (1976) 'Theses on Feuerbach', in F. Engels, *Ludwig Feuerbach and the End of Classical German Philosophy*, Peking, pp. 61–5.

Marx, K. and Engels, F. (1974) *The German Ideology*, London.

Mauss, M. (1979) *Sociology and Psychology: Essays*, London.

Mead, G.H. (1962) *Mind, Self and Society*, Chicago and London, 2 vols.

Mead, M. (1949) *Male and Female: A Study of the Sexes in a Changing World*, New York.

Mechanic, D. and Volkart, E.H. (1961) 'Stress, illness behavior and the sick role', *American Sociological Review*, 26, pp. 51–8.

Mercer, J. (ed.) (1975) *The Other Half: Women in Australian Society*, Harmondsworth.

Merleau-Ponty, M. (1962) *Phenomenology of Perception*, London.

Mészáros, I. (1970) *Marx's Theory of Alienation*, London.

Miller, M.B. (1981) *The Bon Marché: Bourgeois Culture and the Department Store 1869–1920*, Princeton, NJ.

Millett, K. (1977) *Sexual Politics*, London.

Milton, J. (1959) *The Complete Prose Works*, London.

Mogul, S.L. (1980) 'Asceticism in adolescence and anorexia nervosa', *The Psychoanalytic Study of the Child*, 35, pp. 155–75.

Molesworth, W. (1839) *The English Works of Thomas Hobbes*, London.

Morgan, D. (1975) 'Explaining mental illness', *Archives Européennes de Sociologie*, 16, pp. 262–80.

Nicholson, J. (1978) 'Feminae gloriosae: women in the age of Bede', in D. Baker (ed.), *Medieval Women*, Oxford, pp. 15–29.

Nietzsche, F. (1968) *The Will to Power*, New York and London.

Nietzsche, F. (1973) *Beyond Good and Evil*, Harmondsworth.

Nietzsche, F. (1974) *The Gay Science*, New York.

Norton, C.E. (1894) *The Letters of James Russel Lowell*, New York.

Okin, S.M. (1980) *Women in Western Political Thought*, London.

Orbach, S. (1978) *Fat is a Feminist Issue*, New York.

Ortner, S.B. (1974) 'Is female to male as nature is to culture?', in M.A. Rosaldo and L. Lamphere (eds), *Women, Culture and Society*, Stanford, Calif., pp. 67–87.

Ossowska, M. (1971) *Social Determinants of Moral Ideas*, London.

Palmer, R.L. (1979) 'The dietary chaos syndrome – a useful new term', *British Journal of Medical Psychology*, 52, pp. 187–90.

Palmer, R.L. (1980) *Anorexia Nervosa*, Harmondsworth.

Parkin, F. (1979) *Marxism and Class Theory: A Bourgeois Critique*, London.

Parsons, T. (1937) *The Structure of Social Action*, New York.

Parsons, T. (1951) *The Social System*, London.

Parsons, T. (1975) 'The sick role and the role of the physician reconsidered', *Milbank Memorial Fund Quarterly*, 53, pp. 257–78.

Parsons, T. (1977) *Social Systems and the Evolution of Action Theory*, New York.

Pasdermajian, H. (1954) *The Department Store: Its Origins, Evolution and Economics*, London.

Peel, J.D.Y. (1971) *Herbert Spencer: The Evolution of a Sociologist*, London.

Peters, R. (1956) *Hobbes*, Harmondsworth.

Petersen, W. (1979) *Malthus*, London.

Pflanz, M. (1975) 'Relations between social scientists, physicians and medical organizations in health research', *Social Science and Medicine*, 9, pp. 7–13.

Polhemus, T. (ed.) (1978) *Social Aspects of the Human Body*, Harmondsworth.

Pope, L. (1942) *Millhands and Preachers*, New Haven, Conn.

Pospisil, L. (1971) *Anthropology of Law: A Comparative Theory*, New York.

Potts, T.C. (1980) *Conscience in Medieval Philosophy*, Cambridge.

Poulantzas, N. (1973) *Political Power and Social Classes*, London.

Prestwich, M. (1980) *The Three Edwards, War and State in England 1272–1377*, London.

Pullar, P. (1970) *Consuming Passions: Being an Historic Inquiry into Certain English Appetites*, Boston, Mass.

Quaife, G.R. (1979) *Wanton Wenches and Wayward Wives*, London.

Reich, W. (1975) *Reich Speaks of Freud*, Harmondsworth.

Rescher, N. (1979) *Cognitive Systematization: A Systems-theoretic Approach to a Coherentist Theory of Knowledge*, Oxford.

Riesman, D. (1950) *The Lonely Crowd: A Study of the Changing American Character*, New Haven, Conn.

Robertson, D.W. (1980) *Essays in Medieval Culture*, Princeton, NJ.

Rocher, G. (1974) *Talcott Parsons and American Sociology*, London.

Rolleston, H. and Moncrieff, A. (1939) *Diet in Health and Disease*, London.

Rosaldo, M.A. and Lamphere, L. (eds) (1974) *Women, Culture and Society*, Stanford, Calif.

Rose, A.M. (1962) 'A systematic summary of symbolic interaction theory', in Arnold M. Rose (ed.), *Human Behavior and Social Processes: An Interactionist Approach*, London, pp. 3–17.

Rosenburg, C.E. (1979) *Healing and History*, New York.

Rotenstreich, N. (1965) *Basic Problems of Marx's Philosophy*, Indianapolis.

Roth, J. (1962) 'Management bias in social science research', *Human Organization*, 21, pp. 47–50.

Rousseau, J.J. (1960) *Politics and the Arts: Letters to M. d'Alembert on the Theatre*, Glencoe, Ill.

Rousseau, J.J. (1973) *The Social Contract and Discourses*, London.

Rousseau, J.J. (1979) *Reveries of the Solitary Walker*, Harmondsworth.

Rowse, A.L. (1974) *The Case Books of Simon Forman*, London.

Russell, G.F.M. (1970) 'Anorexia nervosa', in J.H. Price (ed.), *Modern Trends in Psychological Medicine*, vol. 2, London.

Sabine, G.H. (1963) *A History of Political Theory*, London.

Sacks, O.W. (1976) *Awakenings*, Harmondsworth.

Sacks, O.W. (1981) *Migraine*, London.

Sartre, J.-P. (1957) *Being and Nothingness*, London.

Scheff, T. (1974) 'The labelling theory of mental illness', *American Sociological Review*, 39, pp. 444–52.

Schmidt, A. (1971) *The Concept of Nature in Marx*, London.

Schochet, G.J. (1975) *Patriarchalism in Political Thought*, Oxford.

Schutz, A. (1962) *Collected Papers*, The Hague.

Schwartz, O. (1949) *The Psychology of Sex*, Harmondsworth.

Scully, D. and Bart, P. (1981) 'A funny thing happened on the way to the orifice: women in gynaecology textbooks', in P. Conrad and R. Kern (eds), *The Sociology of Health and Illness*, New York.

Seguy, J. (1977) 'The Marxist classics and asceticism', *Annual Review of the Social Sciences of Religion*, 1, pp. 94–101.

Seltman, C. (1956) *Women in Antiquity*, London.

Sennett, R. (1974) *The Fall of Public Man*, Cambridge.

Sennett, R. (1980) *Authority*, London.

Sheridan, A. (1980) *Michel Foucault: The Will to Truth*, London.

Shoemaker, S. (1963) *Self-Knowledge and Self-Identity*, New York.

Shorter, E. (1977) *The Making of the Modern Family*, London.

Sigerist, H.E. (1961) *A History of Medicine*, Oxford, 2 vols.

Skultans, V. (1974) *Intimacy and Ritual: A Study of Spiritualism, Mediums and Groups*, London.

Skultans, V. (1979) *English Madness: Ideas on Insanity 1580–1890*, London.

Smart, B. (1982) 'Foucault, sociology and the problem of human agency', *Theory & Society*, 11, pp. 121–41.

Smith, S.R. (1973) 'The London apprentices as seventeenth-century adolescents', *Past and Present*, 61, pp. 94–161.

Smith-Rosenberg, C. (1972) 'The hysterical woman: roles and role conflict in 19th century America', *Social Research*, 39, pp. 652–78.

Smith-Rosenberg, C. (1978) 'Sex as symbol in Victorian purity: an ethnohistorical analysis of Jacksonian America', in J. Demos and S.S. Bocock (eds), *Turning Points: Historical and Sociological Essays on the Family*, Chicago and London, pp. 212–47.

Sohn-Rethel, A. (1978) *Intellectual and Manual Labour: A Critique of Epistemology*, London.

Sontag, S. (1978) *Illness as Metaphor*, New York.

Soper, K. (1981) *On Human Needs: Open and Closed Theories in a Marxist Perspective*, Brighton.

Spiegelberg, H. (1960) *The Phenomenological Movement: A Historical Introduction*, The Hague.

Staples, R. (1982) *Singles in Australian Society*, Melbourne.

Steiner, F. (1956) *Taboo*, London.

Stone, G. (1962) 'Appearance and the self', in A. Rose (ed.), *Human Behavior and Social Processes: An Interactionist Approach*, London, pp. 86–118.

Stone, L. (1979) *The Family, Sex and Marriage in England 1500–1800*, Harmondsworth.

Strauss, A. (1964) *George Herbert Mead on Social Psychology*, Chicago and London.

Strauss, A.L. and Glaser, B.G. (1975) *Chronic Illness and the Quality of Life*, St Louis.

Strauss, R. (1957) 'The nature and status of medical sociology', *American Sociological Review*, 22, pp. 200–4.

Strawson, P.F. (1959) *Individuals: An Essay in Descriptive Metaphysics*, London.

Sturrock, J. (ed.) (1979) *Structuralism and Since: From Lévi-Strauss to Derrida*, Oxford.

Sudnow, D. (1967) *Passing On: The Social Organization of Dying*, Englewood Cliffs, NJ.

de Swaan, A. (1981) 'The politics of agoraphobia', *Theory & Society*, 10, pp. 359–85.

Szasz, T.S. (1974) *Law, Liberty and Psychiatry*, London.

Talbott, J.H. (1964) *Gout*, New York.

Tawney, R.H. (1938) *Religion and the Rise of Capitalism*, Harmondsworth.

Taylor, B. (ed.) (1981) *Perspectives on Paedophilia*, London.

Taylor, F.K. (1979) *The Concepts of Illness, Disease and the Morbus*, Cambridge.

Taylor, G.R. (1953) *Sex in History*, London.

Taylor, I., Walton, P. and Young, J. (1973) *The New Criminology: For a Social Theory of Deviance*, London.

Temkin, O. (1952) 'The elusiveness of Paracelsus', *Bulletin of the History of Medicine*, 6, pp. 201–17.

Temkin, O. (1971) *The Falling Sickness*, Baltimore.

Temkin, O. (1973) *Galenism: Rise and Decline of a Medical Philosophy*, Ithaca, NY.

Temkin, O. (1977) *The Double Face of Janus and Other Essays in the History of Medicine*, Baltimore and London.

Therborn, G. (1980) *The Ideology of Power and the Power of Ideology*, London.

Thomas, K. (1970) 'Anthropology and the study of English witchcraft', in M. Douglas (ed.), *Witchcraft Confessions and Accusations*, London, pp. 47–79.

Thomas, K. (1971) *Religion and the Decline of Magic*, London.

Thompson, E.P. (1963) *The Making of the English Working Class*, London.

Timpanaro, S. (1970) *Sul Materialismo*, Pisa.

Trevor-Roper, H.R. (1967) *Religion, the Reformation and Social Change*, London.

Turner, B.S. (1974) *Weber and Islam: A Critical Study*, London and Boston.

Turner, B.S. (1980) 'The body and religion: towards an alliance of medical sociology and sociology of religion', *Annual Review of the Social Sciences of Religion*, 4, pp. 247–86.

Turner, B.S. (1981) *For Weber: Essays on the Sociology of Fate*, London.

Turner, B.S. (1982a) 'The government of the body: medical regimens and the rationalization of diet', *British Journal of Sociology*, 33, pp. 254–69.

Turner, B.S. (1982b) 'The discourse of diet', *Theory, Culture & Society*, 1, pp. 23–32.

Turner, B.S. (1983) *Religion and Social Theory: A Materialist Perspective*, London.

Turner, B.S. (1984) *Capitalism and Class in the Middle East: Theories of Social Change and Economic Development*, London.

Underwood, E.A. (1977) *Boerhaave's Men at Leyden and After*, Edinburgh.

Vacha, J. (1978) 'Biology and the problem of normality', *Scientia*, 13, pp. 823–46.

de Vaux, R. (1961) *Ancient Israel: Its Life and Institutions*, London.

Veatch, R. (1973) 'The medical model, its nature and problems', *The Hastings Center Studies*, 1, pp. 59–76.

Veatch, R. (1976) *Death, Dying and the Biological Revolution: Our Last Quest for Responsibility*, London.

Veith, I. (1965) *Hysteria: The History of a Disease*, Chicago.

Vermes, G. (1976) *Jesus the Jew*, London.

Vološinov, V. (1973) *Marxism and the Philosophy of Language*, New York.

Wall, F.E. (1946) *The Principles and Practice of Beauty Culture*, New York.

Walsh, M.R. (1977) *Doctors Wanted: No Women Need Apply: Sexual Barriers in the Medical Profession*, New Haven, Conn.

Warnock, M. (1965) *The Philosophy of Sartre*, London.

Wartovsky, M.W. (1977) *Feuerbach*, Cambridge.

Watkins, J.W.N. (1959) 'Historical explanation in the social sciences', in P. Gardiner (ed.), *Theories of History*, Glencoe, Ill., pp. 503–14.

Weber, M. (1961) 'Science as a vocation', in H.H. Gerth and C. Wright Mills (eds), *From Max Weber: Essays in Sociology*, London, pp. 129–56.

Weber, M. (1965) *The Protestant Ethic and the Spirit of Capitalism*, London.

Weber, M. (1966) *The Sociology of Religion*, London.

Weber, M. (1978) *Economy and Society*, Berkeley and Los Angeles, 2 vols.

Weitman, S. (1970) 'Intimacies: notes towards a theory of social inclusion and exclusion', *Archives Européennes de Sociologie*, 11, pp. 348–67.

Wesley, J. (1752) *Primitive Physick, or an Easy and Natural Method of Curing Most Diseases*, London.

Whyte, W.F. (1956) *The Organization Man*, New York.

Williams, R.J. (1963) *Biochemical Individuality*, New York.

Wilson, B. (1966) *Religion in Secular Society: A Sociological Comment*, London.

Wilson, B. (1982) *Religion in Sociological Perspective*, Oxford.

Wirth, L. (1931) 'Clinical sociology', *American Journal of Sociology*, 37, pp. 49–66.

Wisdom, J. (1953) *The Unconscious Origin of Berkeley's Philosophy*, London.

Wolin, S.S. (1961) *Politics and Vision*, London.

Wollheim, R. (1971) *Freud*, London.

Wrong, D. (1961) 'The over-socialized conception of man in modern sociology', *American Sociological Review*, 26, pp. 184–93.

Yuval-Davis, N. (1980) 'The bearers of the collective: women and religious legislation in Israel', *Feminist Review*, 4, pp. 15–27.

Zaner, R.M. (1964) *The Problem of Embodiment: Some Contributions to a Phenomenology of the Body*, The Hague.

Zaretsky, E. (1976) *Capitalism, the Family and Personal Life*, London.

Zola, I.K. (1972) 'Medicine as an institution of social control', *Sociological Review*, 20, pp. 487–504.

# Index